S0-BWO-233

FLORIDA HORSE OWNER'S FIELD GUIDE

Marty Marth

"Marty Marth, an accomplished equine journalist, presents in this text the essential information for responsible horse ownership. With emphasis on Florida, she articulates common situations that daily confront the horse owner, and presents concise and timely information to solve these problems. If you own a horse or are about to purchase one and live in Florida, I recommend this book most highly."

Kirk N. Gelatt, D.M.V.
Dean, College of Veterinary Medicine
University of Florida

"Marty Marth has the rare combination of ability, experience, and enthusiasm to write about horses and Florida's booming equine population. Marth has been a newspaper reporter and editor and has written articles for just about every major U.S. horse publication. Couple with this some 20 years as a rider in Florida and you've got the ingredients of a book that should be read by every Sunshine State horse owner and horse lover."

Bill Landsman
North American Chairman
International Association of
Equestrian Journalists

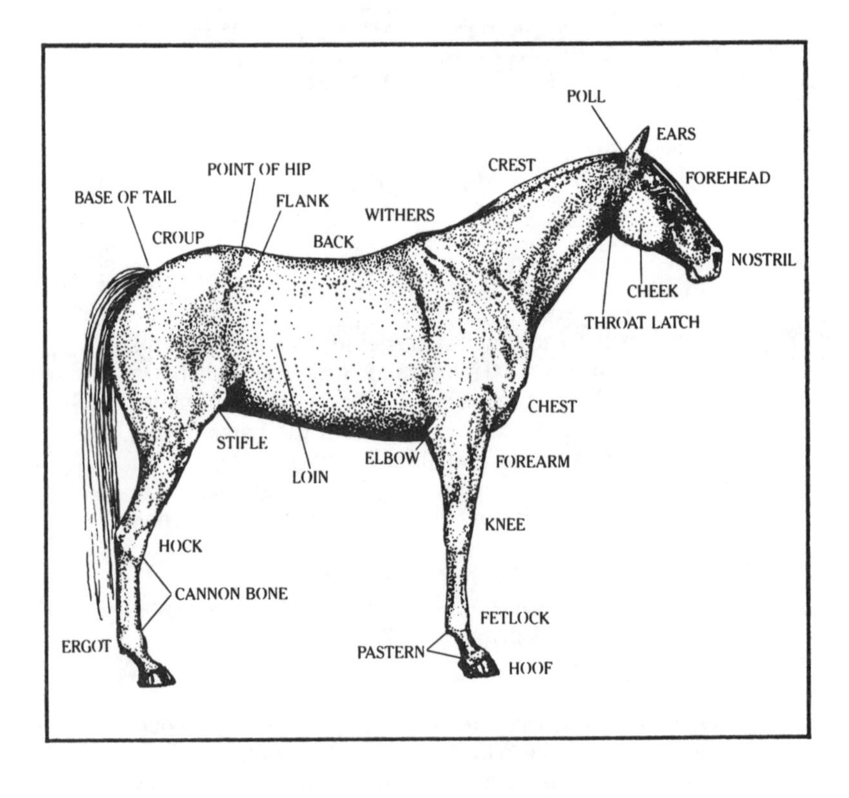
POLL
EARS
CREST
FOREHEAD
POINT OF HIP
BASE OF TAIL
FLANK
WITHERS
CROUP
BACK
NOSTRIL
CHEEK
THROAT LATCH
CHEST
STIFLE
ELBOW
FOREARM
LOIN
KNEE
HOCK
CANNON BONE
FETLOCK
ERGOT
PASTERN
HOOF

FLORIDA HORSE OWNER'S FIELD GUIDE

Marty Marth

Pineapple Press
Englewood, Florida

To Del . . .
And to the beautiful red gypsy horse.
Nobody could have asked for two more patient teachers.

Inquiries should be addressed to Pineapple Press, Inc., P.O. Box 314, Englewood, Florida 33533.

Library of Congress Cataloging-in-Publication Data
Marth, Marty, 1947-
Florida horse owner's field guide.

Bibliography: p.
Includes index.
1. Horses—Florida. 2. Horsemanship—Florida.
I. Title.
SF284.U5M34 1987 636.1'009759 86-30672

10 9 8 7 6 5 4 3 2 1

First edition
Design by Frank Cochrane Associates, Sarasota, Florida
Composition by Lubin Typesetting, Sarasota, Florida
Printed and bound by BookCrafters, Fredericksburg, Virginia

Acknowledgments

The author wishes to thank Dr. Edgar Ott, equine specialist, for his assistance. In addition, pamphlets and bulletins supplied by the Institute of Food and Agricultural Sciences proved invaluable to verify portions of this book's horse care information. Moreover, no writer could produce anything without the helpful persons who contribute their time for personal interviews. The author would like to thank a few of those experts who lent their learned opinions and philosophies over the past decade, including: William Steinkraus, Frank Chapot, Dr. David Meade, Helen Crabtree, Michael Matz, Dr. Ken Vasco, Owen Brumbaugh, Ron Dyer, Dr. Valerie Fadok, and the late Frances Rowe.

Contents

Introduction

Paradise on Earth is to be found in
the arms of a woman . . .
the pages of a book . . .
and on the back of a horse.

— Arabian proverb

Literature refers to the dog as man's best friend. But that is an inaccurate choice. History proves it should be the horse.

Through the centuries the horse was the indispensable companion that worked the land, helped fight the battles, carried the news and fetched the doctor.

Of course, with the development of automobiles, ol' Dobbin found himself put out to pasture. And most historians figured the horse was permanently shelved — relegated to photos in history books.

But the horse has made a dramatic comeback. In fact, the nation's horse population has never been higher. And in Florida, ever-growing numbers of people now seek out horses to ride and drive. The horse is becoming increasingly popular among Floridians — not only for use in recreation and sports — but for breeding businesses and as a status symbol. That groundswell of equine population is verified by a recent upswing in construction of horse-related facilities, particularly in Florida's metropolitan areas.

Palm Beach County's polo industry, with its training facilities and its adjoining pleasure horse barns, cannot keep up with demand; newcomers must stand in line for a stable berth.

So it goes too in Tampa and Orlando. New facilities are popping up overnight — with a trend toward polo, show jumping and dressage. In fact, Florida now is so popular as an equestrian competition site that the prestigious American Grandprix Association schedules a quarter of its 30-competition national tour in the Sunshine State, more than in any other Eastern state.

Moreover, each member of the U. S. Equestrian Team's Show Jumping squad spends more than two months competing in Florida on the Winter Equestrian Festival circuit. This granddaddy of equestrians tours, among the nation's oldest, pays out approximately a half million dollars in prize purses and attracts more than 100,000 spectators.

Similarly, Florida's Thoroughbred, Standardbred and Quarter Horse race tracks entertain more than three million residents and tourists each year and contribute about $20-million to state coffers.

Florida has never produced a reliable equine census. Estimates, based on a flimsy 1971 nose count, tally 200,000 horses in the Sunshine State.

A more accurate estimate may be reflected by the state's Thoroughbred industry, which ranks fourth in the nation. In fact, Florida's major Thoroughbred breeding and training center in the Ocala area ranks in size only behind Lexington, Kentucky.

Yet Thoroughbreds are not the most numerous of Florida's registered equines. Quarter Horses win that honor by comprising 27 percent of the state's equine registrations.

Other breed registries are growing fast, too. Arabian horses, Morgan horses, Paso Finos, American Saddlebreds and many "color breeds" such as Appaloosas and Paint horses also are present in Florida in record numbers.

Because of Florida's rapid equine growth, other facets of the horse industry have grown, too.

To meet medical demands, the state's only equine veterinary college, at the University of Florida, in Gainesville, is undergoing a multi-million-dollar expansion program. The facility offers one of the most innovative and successful neonatal foal units in the country.

And because trail riding is now such a popular recreation for many Floridians, the state has increased funding for a network of bridle paths that criss-cross parklands. And laws have been passed with stiffer penalties for motorists who disturb persons riding horses along Florida's highways.

Explanation for the state's expanding love affair with the horse is simple, basic: A balmy climate permits riding every day of the year. Keeping a horse in Florida is less costly than in other areas, too, because less money is spent on supplemental hay; Florida horses can graze at pasture nearly year around. For the breeding business that means foals can go outside sooner and develop faster. What's more, a unique belt of limestone juts into the state providing North Florida pastures with doses of bone-building calcium. In Central and South Florida, rich mucklands keep grass richly green. As such, then, every horse activity known can be enjoyed in Florida. And each year there are Western and English shows, draft horse pulling contests, three-day eventing competitions, dressage shows, driving and steeplechasing meets and fox hunts.

There are even a number of farmers who still use the horse to plow the fields.

Along with such generous offerings, however, Florida also serves up some unique problems that require special insight. For owning and caring for a horse in Florida's heat and humidity is, in many ways, different than that required in the North.

And that is the purpose of this book: to inform Florida's horse enthusiasts how to get the most pleasure from their four-legged friends in the Sunshine State — the nation's most recognized playground.

Chapter One

Get Horse Smart

A good horse should have three properties of a man, three of a woman, three of a fox, three of a hare and three of an ass.

Of a man: bold, proud and hardy; Of a woman: fair-breasted, fair of hair, and easy to move; Of a fox: a fair tail, short ears, with a good trot; Of a hare: a great eye, a dry head and well running; Of an ass: a big chin, a flat leg and a good hoof.

— Wynkyn de Worde, 1496

Buying a first horse is like getting married. Initial haste affords much opportunity to repent at leisure. Yet too many novice horse buyers plunge into a purchase, seemingly the victims of a stereotype transaction that harkens back to Western movies: Intrepid cowhand lassoes wild stallion and in the next scene placidly plods into the sunset atop his glorious, perfectly-behaved new mount.

Off the silver screen it never happens that way.

Each equine has a unique personality, with distinct likes and dislikes. Therefore the smart, aspiring horse buyer does plenty of legwork before plunking down any cash, not to mention plodding off into the sunset.

The first question a purchaser must ask is: "Why am I buying a horse?"

In Florida, that question can have many answers. The state abounds with a range of equestrian activities, from pleasure riding

"Buyers of horses, particularly horses that will be used in dressage, show jumping and eventing, must be cautious. Buyers often don't bother to examine a horse from a distance, out in a pasture — and fail to get a clear, overall impression of the animal." Dr. Ken Vasco, United States Equestrian Team veterinary advisor.

around the backyard to hobnobbing with the Olympic set at serious hunter-jumper competitions. Usually the novice buyer has the more modest goal — simply to enjoy riding, after work, after school, or on weekends. Far too often, however, this new owner makes a grim discovery: Rather than saddling up the new purchase for years of enjoyment, the owner is instead saddled with an unsuitable animal that is unexpectedly expensive to feed and care for — and surprisingly difficult to resell.

This happens because purchasers ignore the crucial basics of buying the proper animal. Among those basics are size, age and experience of the rider-to-be.

For example, a seven-year-old child is not likely to enter sophisticated dressage competition, so a dressage caliber horse — usually large, skillfully trained, and expensive — is an absurd choice.

The fact is that most first-time horse purchasers are beginning riders. Or parents are buying for beginning riders. As such, they literally have little horse sense when it comes to making the right purchase. Beauty, for instance, is one popular criterion used by new buyers for judging horseflesh. Yet it is probably the poorest. Far more important is a horse's health. Equally significant is conformation. And then there is the matter of sex and size, of age and temperament. Any, and all, can make the difference between an enjoyable hobby or a troublesome responsibility.

Common equine facial markings include (left to right): snip; stripe; bald face; star; star, stripe and snip.

Novice buyers should always ask a trainer or riding instructor to inspect and try out a purchase prospect. Experienced eyes can spot exisiting or potential problems. If an animal gets the pro's nod of approval, then summon an equine veterinarian for a thorough examination of the horse. A veterinarian can detect an animal's more subtle flaws.

Health and conformation are discussed in more detail in later chapters of this book, but some general characteristics are worth mentioning here. For example, a healthy horse has a gleaming coat. Its eyes are clear, its ears alert. Its stance is easy and square on all four feet. And it is approachable, exhibiting a pleasant attitude. Conformation refers to the way a horse's body is assembled. Bowed legs or pigeon toes are a promise of future medical woes because crooked limbs grind away at tendons and joints. Such conformation faults can be serious, eventually rendering a horse unrideable. Others can affect the horse's daily use: A very straight-backed horse may not make a good jumper; an extremely narrow horse might not be well balanced enough for cutting cattle. It is in these areas that trainers and veterinarians lend invaluable advice at time of purchase.

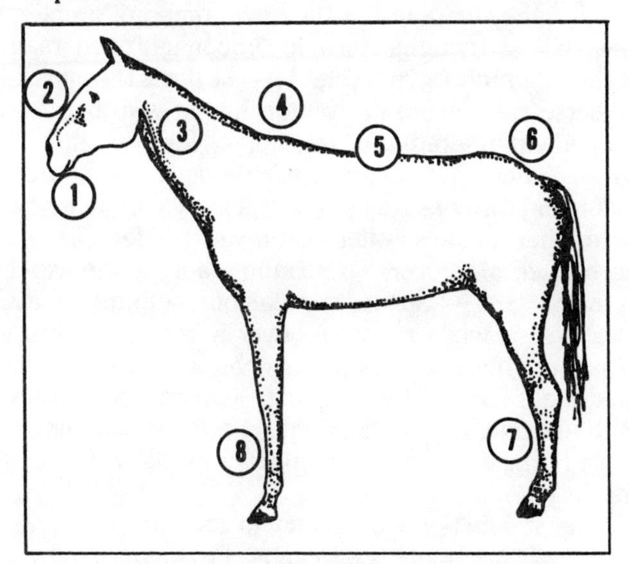

Horse with many conformation flaws, including: 1. parrot mouth; 2. Roman nose; 3. ewe neck; 4. low withers; 5. long-coupled and straight back; 6. short, steep croup; 7. camped-out hind legs; 8. back or behind at the knee.

Another lesson the first-time horse purchaser often learns too late is that the gender, or sex, of a horse can spell the difference between pleasure and problems.

For some unexplainable reason, many novices dream of owning The Black Stallion. Perhaps it's part of an American myth that equates a horse's strength and majesty with virility. Occasionally a properly trained stallion is a feasible purchase. But only occasionally. More often, stallions are hard to handle. Not only are they deceptively strong but they can be, and often are, impossible around a mare who is the object of their affection. Every large Florida barn has at least one trainer or rider who can detail the horror of suddenly "parachuting" off a lusty stallion whose thoughts turned to procreation.

Equine enthusiasts who plan to enter shows also may discover that stallions are unwelcome; that is, some competitions do not permit persons under age eighteen to ride stallions in their classes. And some judges still discriminate in their scoring against women who do compete aboard stallions, believing a stallion is an "inappropriate lady's mount."

Mares are a better choice. They are much less worrisome than stallions, but they can present their own brand of problems. Like stallions, mares' thoughts turn to "moonlight and magnolias" during their monthly estrus cycle. At those times the most kittenish female horse can become a wanton tigress who beckons every male horse within a radius of two miles. However, although they may grow restless and stubborn each month, enamoured mares are seldom as rowdy as stallions. But a mare can be noisy, shrilly yelling to other horses so often that the owner feels like putting a muzzle on her. Moreover, a romantic mare can convince other horses in the barn — particularly stallions — to misbehave.

Nevertheless, while a novice buyer is wise to choose a mare over a stallion, there is yet a better choice. That is a gelding — a castrated male horse. These equine "eunuchs" are easiest of all horses to manage. In general, they accept discipline and training better and make top-notch mounts, especially for beginning owners.

One precaution: Some sellers try to convince first-time buyers that a "proud cut" horse is a gelding. Proud cut is an old-timey equestrian term for a horse that has been castrated but left with excessive scarring, called "proud flesh." Some such castrations are complete, that is, the male hormones don't flow. Other such "cuts" or operations are unsuccessful and the horse behaves like a stallion. For this reason some veterinarians have a saying, "Proud cut is no cut." Always consult an equine practitioner before buying a proud cut horse to make sure you are getting a gelding.

In a similar vein, male horses whose testicles have not descended are a risky purchase. Such horses are known by several terms such as rigs, cryptorchids and monorchids. These horses are usually sterile, but retain their stallion-like urges. Surgery to correct undescended testes is tricky. Some veterinarians will not perform it; those who will do it charge briskly, usually $1,000 or more.

What about size? The horse's height should be coordinated with the size of the intended rider. While this is no life-or-death decision, a rider who is too tall or too heavy for a horse probably will be nearly as uncomfortable as the poor horse.

Horses are measured in terms of "hands" — one hand equalling four inches. The measurement is made from the ground to the withers, a bony point on the horse's neck near where the front of a saddle sits. Some references also call this the point of the shoulders.

Measuring a horse's height often is improperly done. The horse must stand square, preferably on a level, hard surface with weight evenly distributed on all four feet. If front legs are correctly positioned, the elbow (A) should line up with the withers. Few horses are perfectly conformed, which causes some deviance. The horse's head should be held low enough to reveal the highest point of the withers, where measurement should be made. Measuring sticks with a sliding indicator are more accurate than measuring tapes.

Equines who stand 14.2 hands (.2 is a half-hand) are technically ponies; full-sized horses stand 14.3 hands and higher. Some breeds, however, such as Quarter Horses, Arabians, Paso Finos and Morgans, disregard such pony/horse measurement distinctions, preferring to label all their members as "horse" size.

Expected use of a horse determines its preferable height. Animals expected to be jumped over tall fences or to be ridden over cross country courses need to be tall — 16 to 17.2 hands. Pleasure horses, polo ponies or cutting horses can be smaller — 15.2 hands and up — and do their jobs well.

To the dismay of many breed purists, the trend in horse shows has gravitated in recent years toward favoring taller horses. Thus, previously acceptable animals such as the "traditional" Morgan and the "bull dog" Quarter Horse, which usually measure about 14.2 hands, are rather passé. Trendy tanbark thinking is that taller horses have more "show ring presence."

For the novice, however, it is wise to ignore the trends and purchase the size animal that more matches the rider's size, and one that has the necessary height and build to do the job expected of it. Still another consideration is age. Beginners usually fare better with animals over five years of age. Such horses are mature, are usually more settled and, being older, have probably had a fair amount of training.

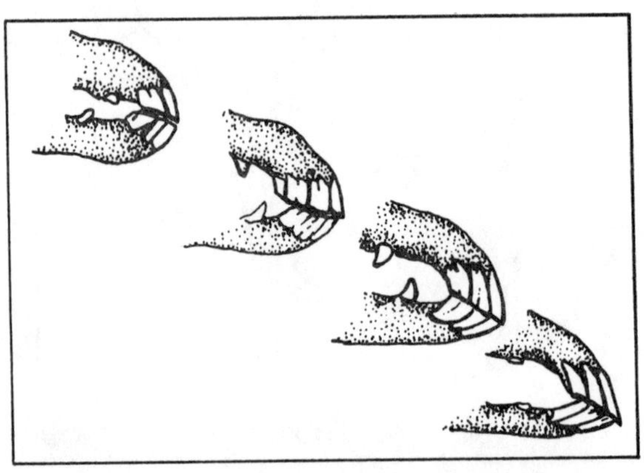

It may be difficult even for experienced horse handlers to tell a horse's precise age by its teeth. Novice prospective purchasers can get a general idea of a horse's age from the jaw profile. From left: A four-year-old horse; a six-year-old horse; an eight-year-old horse; an eighteen-year-old horse. Teeth become longer and more sharply angled with age.

"Attempts to make an old horse look younger are occasionally seen. One trick known as 'gypping' was to paint or dye the graying hair its original color. 'Puffing the glen' was an attempt at filling the depressions above the eye which tend to sink deeper in old age. A more common method of making a horse look younger is 'bishoping'. This involves drilling the incisors of an aged horse and staining the cavity to make an artificial cup, indicating to the novice horseman that the horse is still under 10 years of age." Dr. W. C. McMullan, veterinarian, Texas A & M University.

As with size, the horse's age should not be an overriding consideration. But there is a faulty, generic notion that a horse over age ten is ready for the meat packer. That is not true. In fact, most professional horse people say a horse's prime is clearly around age twelve. At the far end of the age scale are those animals age 20 years or older. It is nearly impossible to determined how many useful years these senior citizens have left. That makes them not the best selection.

As a rule, every "equine" year equals three "human" years. Thus, a ten-year-old horse is roughly equivalent to a thirty-year-old person. Advertisements for "aged" horses usually refer to animals who are eight years or older.

Colts or fillies can be a testy purchase for newcomers because everybody — animal and owner — is inexperienced. A colt is an ungelded male horse under three years of age (except in the Thoroughbred world, which extends coltdom until age four). Colts generally are gelded at twelve months. Young female horses three years old (four with Thoroughbreds) and under are termed fillies.

As in humans, age in horses can be deceptive. How a horse has been used, or treated, over its lifetime can affect its value. Some five-year-old horses may have had a rough upbringing and therefore are fraught with health problems; meantime, a fifteen-year-old, blessed with good genes and careful handling, can be a terrific purchase prospect. What matters most is the amount of training a horse has under its girth.

For example, a "green-broke" (barely trained) eight-year-old horse is a poor option. As horses age, their habits may set like cement; undoing old training, or worse, starting from the beginning, can be a nightmare. This accumulation of knowledge by a horse is often tied in directly with the single most important feature of a horse: its temperament . . . its personality.

Many an otherwise kind ol' Dobbin has arrived at a new barn toting a suitcase that brims with well established bad habits. Stall vices such as kicking, biting, swaying in the stall (weaving) and chomping wood can bring on health problems and shoot feed bills to the moon. A novice buyer can easily spot such habits by visiting the horse in its stall prior to purchase.

The same simple precautions by a buyer can pick up on other equine personality quirks. Some horses, for example, don't mind being ridden but hate to be saddled. One such animal would sink to the ground and lie there until its frustrated owner returned without her saddle; the horse simply preferred to hit the trail in its birthday suit.

Other horses learn to dislike a variety of people. Veterinarians definitely head most horsey hit lists; blacksmiths rank a close second. Then come infants, men-of-any-kind, stall cleaners and feed delivery personnel (not necessarily in that order).

Other horses seemingly come with unwritten contracts that make strict demands: no motorcycles, no airplanes, no puddles. A few prefer end stalls or no stalls. Some dislike automatic waterers. And, would you believe, a few even prefer facing backwards in a horse trailer.

Not all such idiosyncrasies are related to personality; some have to do with health. A horse may be allergic to specific grasses, hays, wormers or grains. Another horse may require roomy digs because it frequently flails around in its stall when it sleeps and gets caught (cast) against the stall wall. These examples are extremes, of course. But they are spelled out here to emphasize that a novice buyer should take time and do some astute observing before making a purchase. Only by visiting a prospect often can the buyer learn whether a particular horse will mesh well with its intended new home.

Closely related to its age and experience is the amount of money a horse is worth. In general, the more training and showing (awards won) the horse has had the more it is worth. Unless a horse owner just likes to brag about the cost of a steed, there is never any excuse to pay more for a horse than it is worth. A smart purchase for the beginner who wants to pleasure ride or to compete in small local shows is a "grade" horse. Grade animals are the unregistered, non-pedigreed "mutts" of the horse world. They may lack the quality of registered animals, may command a lower resale value (although not always) than a registered horse, and may be unsuited for the higher-rated hunter shows — but, for the novice, their advantages far outweigh such few disadvantages. Grade horses can compete extremely well in performance classes such as eventing, jumping, dressage, and barrel racing. These activities alone provide a lifetime of pleasure and competition. Then there is the cost. Grade horses almost always cost less than registered animals.

A horse's general attitude is often revealed by the horse's head position and how it places its ears. The top horse is alert but calm; ears flattened at the handler's approach (bottom left) can indicate fear or a nasty disposition — possibly a horse that will bite; head held high (bottom right) with ears pricked signals a horse that is hearing or seeing something that it doesn't understand — an animal about to shy.

Moreover, some so-called grade horses are actually purebreds but lack pedigree papers. So the buyer with an eye for horseflesh may find a real gem. Many examples exist of fine Thoroughbreds or Arabians who, while lacking papers and therefore considered to be grade horses, perform well and win at shows. In most instances the cost of these "finds" was typical for grade horses — $400 or $500. And because they performed successfully, these grade animals' resale value was nine or ten times their purchase price. As for the price of a registered or purebred horse, the limit could challenge even the most generous pocketbook.

Some purebreds are available for about $1,000, but these animals are usually "culls" — less desirable because of mental or physical defects. Most well-bred registered horses are priced from $3,000 on up. And registered or purebred horses who have been widely shown and who have won prizes in bigtime shows usually begin at $10,000 and extend into six-digit figures. Of all horse purchases, none comes closer to gambling than buying breeding stock. Value of a broodmare or stallion is usually based on the amount of "black type" — ancestral standouts — the animal has in its pedigree. Yet no amount of ancestry guarantees that a particular horse will present the world with sterling babies. In fact, it sometimes takes generations before outstanding features of a stallion or mare become evident. If the novice buyer even entertains the notion of purchasing breeding stock, a skilled professional should be in attendance. In nearly all cases, however, the professional is to be brought in for the finale to the act of choosing a horse. Decisions as to what type of horse, age, sex, size and the like must made by the prospective purchaser. Once that is done, the buyer is ready to begin the hunt for that "perfect" horse.

Chapter Two

Making A Savvy Buy

In the choice of a horse and wife,
A man must please himself . . .
— Arabian proverb

Fortunately for Florida's horse hunters, the Sunshine State offers plentiful prospecting. Few states have more equines, more equestrian activities or more purchase opportunities.

Basically, there are two ways to purchase a horse: via a mass sale such as an auction, or from private individuals. The latter may breed or train horses, or wish to sell their horse because they are moving, are getting out of the hobby or have outgrown their horse.

Auctions come in different forms. There is the weekly or monthly variety held in livestock pavilions or rural auction houses. There are the large breed auctions, such as Thoroughbred or Quarter Horse auctions, that are held only a few times a year. And there are the disbursement sale — a breeding or training barn "cleans house," selling off its stock because the barn is overcrowded or is going out of business. (Some disbursements are not auction-type sales.)

Shakiest of the auctions for novice buyers are the weekly sales. Although interesting to attend, they present more pitfalls than good buys for beginners. One problem is purchasers have no opportunity to see the auctioned animals under "normal" conditions, that is, in home stalls or pastures. Also, buyers usually cannot ride the animals prior to the sale. Still worse, the medical and breeding histories of auctioned horses in some instances are phony or nonexistent. Most auction sales are final; an unsatisfactory selection may not be returned.

Unfortunately for both the horse and the bidder, an animal brought into the sales ring at this type of auction may be a different creature than it was upon arrival on the grounds. Prior to sales time, the horse may be kept in cramped, unsanitary "feed lot" conditions. Healthy animals mingle with sick or vicious ones. And the decent stock often is whipped or spurred into the sales arena with the result that the animals are "equalized," i.e., frightened good horses may misbehave, frightened unsuitable horses may act abnormally subdued. Morever, the animals' briskly pumping natural adrenalin, or a drug administered prior to the sale, may mask medical problems.

In truth, these weekly horses auctions usually offer grade horses that either nobody wants or that owners cannot sell by other means. Sadly, the most vigorous bidders at these small auctions are representatives from meat (dog food) companies.

Readers should not confuse these small grade auctions with the classy breed auctions that are held only once, twice or three times a year, say, in Ocala.

Whereas the weekly sales offer mostly rejects, the prestigious breed auctions concern themselves with some of Florida's most promising equines. Consequently, rarely at a breed auction is a horse sold for less than $2,000. Some sell for millions of dollars. In either case, the price tag usually is too rich for a beginner's budget.

Best source for a novice horse purchaser is either a reputable breeder or an individual owner. And often the best "inside scoops" on quality sale animals come from the area's equine veterinarians or farriers. Other good sources include classified advertisements in daily newspapers, horse publications and local horse club newsletters. Occasionally, too, bulletin boards posted at horse shows list excellent buys.

Yet another likely source are the many training and boarding facilities around the state. It is rare to visit such a barn without seeing at least one horse for sale.

Advertisements are good leads but because of space limitations they usually tell little that a horse buyer needs to know. These leads can be taken only as that, leads, and require a phone call and a barn visit for more specific information. Some ads do contain words or phrases, however, that can save the buyer's time. For example, the beginner should be leery and back off ads that say "for experienced riders" or "a fast barrel racer" or an "ex-racehorse." The last thing a novice rider needs is a stunt man's mount or a race track reject. Thoroughbreds, Quarter Horses and trotting horses (Standardbreds) that come off the track can be risky purchases. They are usually trained quickly to save expenses and too often are kept on the track long after any injury has passed the "salvage" point. Some ex-racehorses are not only physical wrecks but also have mental problems. They were bred and trained to run, so it can be a heady challenge for a beginner to convince an ex-racer that it no longer must gallop six furlongs every time it is saddled up.

Standardbreds can present their own unique problems. These horses are urged to perform at their best natural gait, meaning trotting or pacing. Cantering is a no-no and some of these horses were cruelly disciplined whenever they did break into a canter. The result: some Standardbreds never will canter in a relaxed, natural manner.

Pacing horses (left) move lateral legs, those on the same side of body, forward at the same time. Trotting horses, (right) move diagonal legs together.

Advertisements usually are placed by sincere, honest sellers who hope to find good homes for their horses, but enough unscrupulous individuals have been out there since the Roman era to give the term "horse trader" a bad connotation.

An unscrupulous seller today will hide a horse's unruly behavior with tranquilizers. It helps to know some telltale signals of a doped horse. For example, a sound dose of "Tee" can cause a male horse to droop its penis. And a tranquilized horse will often look sleepy and not stand squarely on all four feet; instead, it leans on three legs, propping a hind leg ahead. Likewise, a sedated horse's ears will not prick alertly. The animal also may hang its head and act generally unresponsive.

Besides tranquilizers, other medications such as painkillers can mask imperfections. Thus an unsound animal may travel like a champ as long as it is on such a medication. Once it wears off, the horse's unsoundness becomes obvious.

Lameness is one such unsoundness. A lame horse is a useless horse. Unfortunately, lameness is sometimes hard to detect; and once it is detected it requires a thorough knowledge of horses to determine whether that particular lameness is temporary or permanent.

Therefore it cannot be stated too strongly that purchasers have a horse examined by an equine veterinarian before coming to a decision. A vet, for example, usually administers flexion tests in which certain limbs of the horse are stressed after which the animal is trotted to see it if is lame. A vet also will study the horse from all angles while it is being led around at the trot for it is only at the trot, not the walk, that many lamenesses appear. In some instances, where the veterinarian is suspicious, he will offer to take X-rays. These boost the exam cost, of course, and the decision to have X-rays taken should depend on how badly the purchaser wants that particular animal.

Cost for a veterinary purchase examination will vary. A $15 to $30 "trip charge" is average, with the remaining fee based on how much effort and time the doctor spends examining the horse. If X-rays are taken and detailed lameness procedures need to be employed, some exams can cost $100 or more. Initially, such a charge can pinch the pocket, but buying a horse that a few days or weeks later turns up lame may require a far greater outlay of veterinarian bills if the horse ever is to be useful.

Interestingly, some veterinarians will give the okay to an animal that another vet may refuse to pass. If a purchaser is not satisfied with a veterinarian's decision, he or she should always (if the checkbook can take it) summon another opinion.

Beckoning a veterinarian to check out every equine prospect can grow prohibitively expensive. For this reason, most buyers initiate their own "weeding" process, leaving only the most desirable horses to be vet checked. Separating the good prospects from the bad ones is not difficult for the novice buyer but does require some fundamental education to distinguish between mere equine blemishes and more crucial unsoundnesses. Some "unsoundnesses" make their presence known almost immediately. So do some "blemishes." But a blemish does not mean a horse is unsound. For example, a horse may be scarred by a rope burn or a wire cut; such a blemish may not affect a horse's usefulness. The same is true for floppy ears or a scarred nose. It is useful, however, for the purchaser to ask how the horse acquired its blemish. The answer may be revealing. A horse may have a scarred nose from rearing in a trailer, or floppy ears from improper ear twitching, a technique that distracts a horse's attention while its being clipped or otherwise tended. In these cases, the purchaser, by asking about the blemishes, has gained some useful knowledge. The scarred-nose horse may be impossible to trailer because of its previous bad experience, or the floppy-eared animal may be forever edgy about clippers because it was twitched to the point of injury. It pays to ask questions.

Nevertheless, blemishes should not be confused with unsoundnesses, some of which are listed below. Each example, especially if blatant, should raise a caution flag to the buyer that it might be wise to keep shopping.

Sight defects. Horses function best with two eyes. Seldom is a one-eyed horse usable; in fact, it is usually dangerous because it may be a chronic stumbler.

Wind defects. A horse whose breathing is labored, or whistling, may have emphysema or some other ailment that qualifies it for the "injured reserved" list.

Limb defects. Any animal that cannot move at all or without marked stiffness deserves probable rejection.

Disease. Diarrhea, a chronic cough, and a thick, infectious-looking nasal discharge would indicate a sick animal.

Conformation. As mentioned in Chapter One, the way a horse is assembled is important to its use, its performance, its health and certainly to the comfort and pleasure of the rider. While there is plenty of room for personal taste, a few fundamental conformation ideals should be sought.

The buyer will, of course, want a horse that is pleasing to the eye. Joints should be smooth so that each body part appears to "flowingly attach" to the next. The horse's legs should look smooth, showing no large swollen areas or unnatural bony protrusions. Pay attention to shoulder and hindquarter angles. For example, the preferred horse's shoulder angle should approximate a 45-degree angle measured from near the front legs to the withers. Similarly, reasonably prominent withers are always desirable for it is easier to anchor a saddle on a well withered horse. (Saddles slide more on a "muttoned-withered" horse, meaning a horse with low withers.) While no precise angle measurement is afforded a horse's hindquarters, some slant of the quarters from the hip to the buttocks point (the croup area) should be present.

Such angles are important because "straight" horses — those that tend to have straight shoulders and a level croup — more often have limited shoulder and hind end mobility for activities such as dressage and jumping.

Moreover, "straight" horses tend to be straight everywhere. A straight-shouldered horse will have straight pasterns, for example, and such "upright" pasterns signal a horse with gaits that are "trappy" or choppy. It makes for an uncomfortable ride. Pasterns, like shoulders, should angle some 45 degrees. They also should be smooth, for odd lumps and bumps on pasterns may reveal side bone or ring bone, which can cause lameness.

Perhaps no part of a horse is as important as its foot. Composed of numerous tiny and fragile bones, a horse's foot frequently is the focal point of recurring medical problems. Never buy a horse without checking each foot or hoof.

Look for feet that are nicely shaped, fairly round, and proportioned to the horse's size. The back or heel of the foot should be wide and deep. On the bottom of the hoof, when lifted, is the sole; it should be clean with few cracks and crevices. Near the heel the V-shaped cleft, called the frog, should look pliable and well developed. Novices too often ignore the frog, yet it is the "foot pump" that circulates much needed body fluids through the feet and legs. A puny frog may indicate the horse is neglected and either has been lame, or may become lame.

A horny substance, called periople, covers the hoof's outside wall. It should be naturally glossy and smooth. The hoof outline should be straight, not ringed with grooves. Such grooving, combined with a hoof that appears sunken in the middle, is a sign the horse may have foundered (suffered a high temperature) at one time. The ailment occasionally leaves the hoof bones pivoted out of place, a forerunner of lameness.

These recommended "weeding out" processes may seem technical, but after a buyer has inspected a few horses such eyeball examinations become nearly second-nature. The purpose in mentioning them is to encourage the buyer to learn how to judge a horse's major good and bad points. A veterinarian can be relied on to spot more picky problems.

There is no better place to inspect a horse, to make the initial examination, to learn a horse's habits, than at the horse's home — its stall. It is loaded with clues.

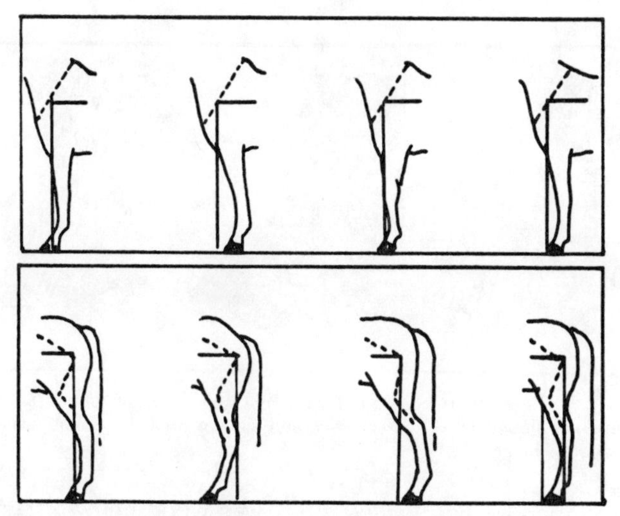

Foreleg conformation problems viewed from the side are (from left): desirable, too far under the body, over in the knees, calf-kneed.

Hindquarter faults seen from the side are (from left): desirable, sickle-hocked, legs too far back, overly straight hock.

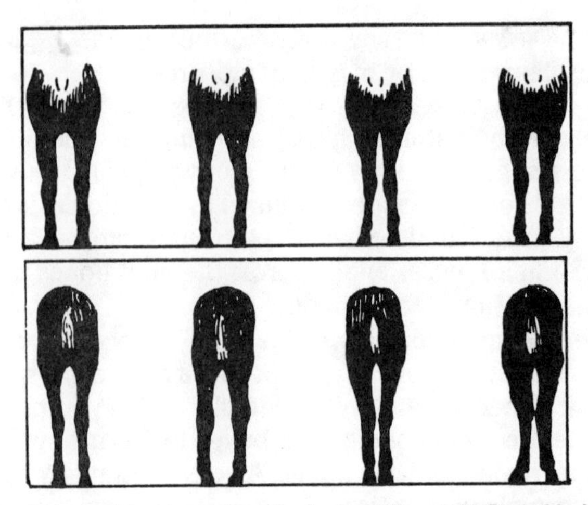

Conformation problems seen from the front are (from left): Desirable, bow-legged, knock-kneed, pidgeon-toed. Hindquarter faults seen from behind are (from left): Desirable, bow-legged, too close, cow-hocked.

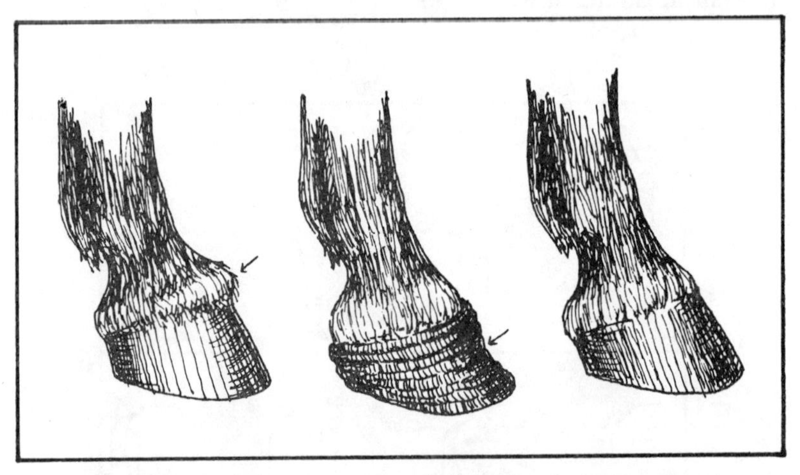

THE FOOT: (Right to left) Normal foot. Foot that shows symptoms of past laminitis (founder) — note grooves and sunken area. Foot that reveals ringbone bulge.

Mentioned earlier, for example, was wood chewing. Look for marks on the stall door or inside the stall that reveal that the horse chews wood or kicks. Merely chewing wood is no grave vice, but a chewer may be a "cribber," that is, it clamps its teeth on a piece of wood and sucks air. The habit is unhealthy and difficult to break. Sellers often try to hide the fact a horse is a "cribber." They may have tried to break the habit. If so, a telltale sign may appear at where the horse's neck joins the head: If a wide band of hair is missing, suspect that the horse regularly wears a "cribbing strap."

Always check the stall floor. If a path is worn down in it, the horse probably is a pacer or weaver. Both vices use up an animal's energy and can indicate a nervous personality.

If the stalled horse is fidgety, whinnies often to its neighbors and seems generally upset, ask the seller if the animal has just been moved to that stall. Moving a horse is a trick some sellers use to hide stall evidence of an animal's bad habits.

If the stall contains a fan, ask why. Many Florida horse barns use fans (normally on the ceiling) to keep flies away, but a fan could indicate a more serious problem than insects. It may be a sign that the horse does not sweat. In Florida, a non-sweating horse must be kept in a cool place because his temperature-regulating sweat does not flow properly. And a non-sweating horse in Florida's summer

Telltale wear signs on stall surfaces and interiors may indicate problems to a prospective horse purchaser. This stall's occupant may be a digger (cavity under door), may be a wood chewer (worn notches on top boards), may kick (hole in front board), and may not sweat (fan installed on wall).

is fairly useless. If the animal cannot sweat it cannot keep its internal system cool. The result is its temperature soars during a normal riding session, leading to heat prostration.

The most telling sign of a non-sweater is an air conditioned stall. Any horse kept in such quarters should be immediately scratched off a buyer's list.

Whenever horse hunting, visit the prospect more than once. It is an excellent appraisal technique. The first visit should be by appointment, of course. But if the horse is worth another look, "pop in" unannounced another day to see the animal again. If on the unannounced visit the horse's owner or handler refuses to let you see the horse, or delays your examination for more than ten minutes, be suspicious.

It would behoove a novice shopper to learn how to properly lead a horse at the walk and at the trot before starting off to barns for examinations. With this basic knowledge, a buyer, by walking and trotting the animal, can get a "feel" for the horse, determining whether it is friendly and obedient. After walking or trotting the horse, have the owner or handler walk and trot the animal, too, and on firm, level ground. The buyer can get a good look at how the animal moves. If the horse travels unevenly or either of its shoulders or hindquarters seems lower than the other, suspect that some anatomical feature is out of kilter.

Intelligent questions also will produce valuable information. The savvy shopper may not initially understand why some of the following queries are valuable, but as the months pass, the new owner will be happy to have asked questions such as:

Does the horse longe? That is, does it work at the end of a 30-foot line? This basic training technique is not an absolute necessity, but it is desirable.

How much does the horse eat and drink?

Has the horse ever colicked? Many mature horses have colicked at some time and honest sellers will say so. A past colic is no reason to dismiss an animal.

Does the horse load and ride nicely in a trailer?

Does the horse trail ride?

Does the horse tie? It's helpful if a horse has learned to stand in cross ties. Some horses ground tie, that is, they stand perfectly still when the lead line is dropped.

How often is the horse wormed? Well-tended horses should be on a regular worming program. If the seller admits the horse has never been wormed or is wormed only once a year, be wary. Poor worming practices may foretell future bouts of colic.

Is the horse pregnant? Many buyers neglect asking this question and find a little surprise has arrived months later. If funds are limited, this bundle of joy may be unwelcome for two equines cannot live as cheaply as one.

How many times has the horse been sold? A horse that has been owned many times may be wearing a subtle sign that says, "I'm not so hot once you get to know me."

Has the horse ever been in an accident?

Once you and your veterinarian are satisfied with a horse, it's time to yank out the checkbook. Keep the sale simple. Buy a horse outright and sell a horse outright. Too many novices get trapped on a trade-in "merry-go-round" when the seller says, "If this horse does not work out, I'll take it back on a trade for another of my horses." This sounds tempting, but such deals are rarely even trades. After three such typical trades the bookkeeping looks poor:

Purchased Horse A (worth $500) . . . paid $500. Later traded in Horse A (still worth $500) plus $400 cash for Horse B (also worth $500). A few months later traded in Horse B (still worth $500) plus $300 cash for Horse C. Horse C actually cost the unwary horse purchaser a total of $1,200 (plus whatever horses A and B were fed). Yet it's always surprising to hear horse C's owner brag that the animal is "a $500 horse that I got for $300." The wily trader, meanwhile, not only has pocketed $1,200 for the three horses, but has resold horses A and B for an additional $1,000 — thus banking $2,200 for $1,500 worth of horseflesh.

Sales paperwork that should accompany any equine includes the sales contract, registration papers if applicable, and in Florida, a negative Coggins certificate (see page 111).

Some purchasers like to have an attorney draw up a sales contract. This is fine for five-digit purchases or complicated deals. But all that is necessary for most sales is a precise statement of who is getting what. The contract should include the date, the purchaser's name, the amount paid, the horse's description (including age), and both party's signatures. Of course, any special purchase terms such as a return or refund agreement, installment payments, or special equipment that comes with the horse should be listed in the contract. If the sale is subject to a veterinary okay, that also should be spelled out along with any deadlines by which the vet check must be completed and the deal either finalized or nullified.

Another desirable "extra" to get from the seller is any available past medical receipts the seller has for the horse. This aids the buyer to schedule routine inoculations and wormings, especially if the horse will be moved to another locale in Florida. If the horse is not being moved, the new owner should plan to use the veterinarian and blacksmith who is most familiar with the horse, at least for a while.

When purchasing a grade animal, ask about its parents. Certainly this data is not critical, but it can be useful if the horse is: a) a purebred animal that lacks official papers; b) a mare that eventually may be bred; c) such a wonderful animal that another of that particular breed combination may be desired.

Once the horse is purchased, consider insurance. (An unusually valuable animal should be insured as soon as the purchase is made.) Some insurance firms will protect against injury caused by a horse as part of a homeowner's policy. Most experts agree, however, that equine insurance coverage is best obtained from reputable equine insurance specialists.

Chapter Three

Breeds: An Equine Smorgasbord

. . . they bred such horses in Virginia then. Horses that were remembered after death And buried not so far from Christian ground.

— Stephen Vincent Benet

Purebred horses abound in Florida. While they often sport high price tags, they usually also give the purchaser a breeding or show quality animal. In fact, the biggest dilemma for a prospective buyer of a purebred is choosing from dozens of breeds. Each breed boasts a special use, a unique temperament, a special beauty or color. Then there is that favorite sales buzzword — "versatility." Virtually all breeds, through their owners, claim to offer the only "truly versatile" horses — animals that can do it all — from working cattle to jumping Olympic-level fences. Such claims are usually exaggerations, of course. A draft horse, for example, isn't about to execute a point-scoring pirouette.

The best way to determine which breed is best for specific uses is to attend shows and watch the horses in action — and ask questions. While some basics can be gleaned from reading, nothing beats firsthand observation.

Meanwhile, the following breed summary can help pare down the selection task.

MAJOR PERFORMANCE AND CONFORMATION BREEDS IN FLORIDA

Let's begin with those major breeds found in Florida that can be used in horse shows because they meet certain conformation or appearance requirements. Bear in mind these are breeds that are not considered "color" breeds, that is, they are not bred for their color; that category is covered later.

American Saddlebred

The American Saddlebred is also called the American Saddle Horse breed. As with other breeds such as the Morgan, the American Saddlebred is promoted by its owners as "the truly American breed . . . developed by pioneer horsemen."

American Saddlebreds trace bloodlines back to Thoroughbreds, Standardbreds, Morgans and Arabian horses. Blended in, too, is blood of "native ambling mares," probably hardy horses of no particular breeding that added a propensity toward the ambling walk — a gait in which the left fore leg and left hind leg move together and the right fore and hind do the same.

American Saddlebred

During the 1700s, Saddlebreds originally were bred as general utility horses. They were used for riding and pulling wagons. Today the breed, which stands 15 to 16 hands high, has come full circle. Now known for its elegance, the Saddlebred today is most often used as a show horse and is exhibited as either "three-gaited" or "five-gaited." Three-gaited Saddlebreds exhibit at the walk, trot and canter. A high-stepping brilliance is valued and is achieved through training and the occasional use of weighted, built-up shoes. Five-gaited Saddlebreds also must display a high degree of animation at the walk, trot and canter but additionally they perform two other gaits. One is called the Slow Gait, while the second, the Rack, a speedy cousin to the Slow Gait, is a fast ultra smooth running walk. Saddlebred horses are specifically groomed according to their use. For example, three-gaited horses wear their manes and forelocks "roached" (completely shaved off), while five-gaited horses wear long manes. Both styles require long tails that may be "nicked," although this practice is more common to three-gaited horses. Nicking is a surgical procedure in which retractor muscles near the tail base are clipped, thereby preventing the horse from clamping its tail down over its rump. As a result, the tail is always carried high in a sort of bantering position. Five-gaited Saddlebreds may wear a device that holds the tail up so nicking is not necessary. American Saddlebreds are seldom chosen for hunting, jumping or dressage activities. Breeding trends may explain this phenomenon inasmuch as Saddlebreds are prized for their straight backs and having a neck that is set high on the body. While attractive to look at and smooth to ride, horses with such traits are not particularly useful for jumping fences or riding dressage tests. The breed, however, makes a suitable pleasure or trail riding mount. Unfortunately, because some Saddlebred training practices emphasize excitability, the breed has acquired a reputation for being slightly flighty to handle. In defense, breed owners point out that every breed includes a few individuals who are highly strung. Nevertheless, such excitability looks good in show business and it is for the show ring that the American Saddlebred remains most highly treasured. Give these elegant creatures a glowing spotlight, a top-hatted rider, and some toe-tapping music and the Saddlebred is one of horsedom's most dazzling breeds.

Arabian

Perhaps no breed is as well known or as romanticized as the Arabian. Such notoriety may stem from the mysticism of Arabian tales. No matter, for the notion of the handsome desert sheik — or his feminine counterpart — aboard a fiery Arab steed, is fondly fostered by Arabian horse breeders and their well-oiled public relations machinery.

Nobody knows for certain just when the breed began. Most estimates say about 5,000 years ago. Bedouin tribes are credited with first taming and training the breed, although the horses may have been driven rather than ridden. Whatever their use, the Arabian became a valued commodity and the breeding records (matriarchal rather than patriarchal lines) were memorized by tribal entrepreneurs. The breed is thought to have been first brought to the United States in about 1760. Horses brought to Florida by Spanish explorer Hernando de Soto in the 1500s are believed to have contained Arabian blood. The breed has become especially popular in the state since the 1960s.

Arabian Horse

Any prospective buyer soon discovers "two" strains of Arabian horses dominate the pedigreed breeding market: Polish and Egyptian. Photographs of the two types defy great visible differences although experts of each type contend they can spot the subtleties of variation.

A currently desirable Arabian horse stands 14.1 to 15.1 hands high, has a level back, a long, curving neck and a high-set tail.

Breeders of Arabians also promote their horses as among the most versatile. A typical promotion brochure says, "Members of this breed, properly selected for the job, can be trained to do anything any specialist breed can do." The same brochure qualifies that statement and admits the Arabian horse "cannot run as fast as the Thoroughbred, or trot as high and fast as a Saddlebred, but he averages out very well."

Indeed, the Arabian horse does average out very well. The breed is intelligent, compact, maintains well on moderate amounts of feed, and has proven to be excellent in a variety of activities, especially endurance events. Virtually all other light horse breeds (heavy breeds being generally draft or work horses), including the American Throughbred, trace to Arabian origins. Arabian horses are one of Florida's fastest growing breeds. In recent years, Ocala, the state's Arabian Mecca, has enjoyed a five-fold increase in Arabian farms. This popularity is due primarily to an active breeding society and to an ongoing push to expand the Arabian horse racing industry.

Prospective Arabian purchasers should be aware that this popular breed suffers minor limitations in the show ring. Open shows that offer hunter, Western or Park Pleasure classes, for example, often favor breeds that traditionally perform in such classes, those breeds being Thoroughbreds, Quarter Horses and Saddlebreds, respectively. However, the Arabian is becoming such a favorite in Florida that numerous all-Arabian shows are held weekly in all areas, providing a wealth of showing fun and experience for breed owners.

Morgan

Morgan horses have an especially loyal following among equestrians. The breed reportedly began in 1789, the same year George Washington became president.

That year a bay colt of unknown parentage was born in West Springfield, Massachusetts. The story is told that Justin Morgan, a Vermont schoolmaster, walked to West Springfield to collect a debt owed him by the colt's owner. Morgan ended up walking back to Vermont with, in lieu of cash, the young bay colt and a second adult horse. Morgan sold the mature horse, say historians, but no buyer wanted the small bay colt named Figure. "Too small for the rugged Vermont Green Mountains," everybody said. Over the years Figure outworked most of the other area breeds, and the breed's historians reported Figure passed on to his offspring his amazing endurance, his sturdy, short-coupled build and his intelligence.

Tradition in the 18th century dictated a horse become known by its owner's name; thus, the horse Figure became the Morgan horse. Throughout U.S. history, particularly during the Civil War, the breed has proven itself a valuable mount. Company H of the Fifth New York consisted of 108 Black Hawk Morgans, seven of which survived the North-South struggle. And the Black Hawk Morgan ridden by Union Gen. Philip Sheridan has been preserved and is on display in the Smithsonian Institution's American History Museum in Washington, D. C. In sports, the Morgan became a popular trotting horse. The breed's fans maintain that Morgans are the horses depicted in Currier and Ives' racing scene prints.

As race breeding progressed, however, the purity of the Morgan breed began diluting. Fortunately for the breed it found an ally in Col. Jospeh Battell. He became fascinated with the Morgan in the 1870s, and for 15 years Battell researched Morgan pedigrees, even publishing the first Morgan Horse Register in 1894. In 1906 Battell donated his Middlebury, Vermont, farm to the U. S. Government to foster the breeding and restoration of Morgan horses. For 45 years (1906-1951) the U. S. Government established and operated the United States Morgan Horse Farm. There, the Morgan horse's stamina and endurance was tested in order to identify outstanding individuals. The best stallions were used as sires for cavalry mounts. In 1952, the Government gave the breeding operation to the University of Vermont. Some Morgan historians estimate 85 percent of today's Morgan horses trace their bloodlines back to the Vermont farm. Twenty-four years later, in 1976, the Morgan was named the official "Bicentennial Horse."

Despite its talent and beauty, the Morgan breed, like the Arabian, usually does not fare well in the show ring against Thoroughbreds in hunter classes or against Quarter Horses in Western stock classes. The Morgan shines best in versatility or pleasure classes. Outside the show ring the Morgan is adept at every type of equine work and pleasure.

Morgan horses figure in the blood backgrounds of the American Saddlebred horse and the Tennessee Walking Horse. Yet Morgan breeders today seem discontent with their stock as they attempt to breed taller Morgans. Thus, the once typically compact 14.2-hand breed gradually is being replaced with taller breed examples, some towering up to 16 hands.

Quarter Horse

Florida statistics reveal that no breed in the Sunshine State is as popular as the Quarter Horse. Of the state's total equine population, more than 27 percent — almost one-third — are registered Quarter Horses. Most members of the breed live in the state's central and southern areas such as Dade, Orange, Palm Beach, Hillsborough and Polk counties.

As with most other American breeds, the Quarter Horse traces back to this nation's Colonial days — to Indian or Spanish horses that were crossbred with English imports. Colonial Americans delighted in racing their horses. Traditional oval tracks were not yet common so speed contests often were held on short and straight 440-yard clearings. It did not take long for the racing enthusiasts to discover that their crossbred horses could polish off the quarter-mile distance in a flash of blinding speed. The breed became known as American Quarter Running Horses.

Quarter Horse

But Running Horses' popularity waned in the mid and late 1700s. Arriving settlers were more affluent, more interested in raising Thoroughbred racehorses to race on oval tracks. Smallish (14.2-hand) Running Horses were relegated to the farms. The Running Horse accompanied settlers who headed south and west. Short distance match races remained popular with them and they took great pride in their fleet and sturdy quarter-milers. When the travelers reached Texas, they were influenced by Spanish-style Mexican horsemen who used horned saddles and spade bits on their horses. Mexican riders also efficiently roped cattle with long *reatas* (lariats). The sturdy Running Horse soon showed a high degree of "cow sense" — ability to separate or "cut" one cow from the herd and to keep the lariat taut while a rider performed necessary chores on the roped cow.

By 1941, the Running Horse was officially named American Quarter Horse, its breed registry was started and its own association was formed.

Popularity of the Quarter Horse has soared over the years. But as uses of the breed expanded, demands increased and owners searched and bred for greater breed variety. Today, the short, blocky "bulldog" type Quarter Horse of decades past is rare; the taller (15.2 hands and up) specimen is more usual.

The breed is a Florida favorite because of its calm temperament, good size, its square and sturdy build. Quarter Horses come in 13 recognized solid colors and in a variety of types that range from the previously mentioned solid compact horse to the taller, more Thoroughbred-type. As a result the breed can be successfully campaigned in not only Western pleasure and speed classes, but in English Hunt Seat classes and over grand prix level jumps. Appendix-registered Quarter Horses are especially popular for the hunter ring. This branch of the Quarter Horse registry is for equines who have one registered Thoroughbred parent and one registered Quarter Horse parent. Quarter Horses are usually not recommended for English Park Pleasure classes nor for dressage competitions. The breed's famously smooth "daisy clipping" movement often lacks the necessary animation and exaggerated stride to do well in either. The breed also has been criticized because of a tendency by owners to exhibit, at halter, overweight animals (excessive weight covers up conformation sins). To compound health problems, Quarter Horse fanciers have bred horses to achieve overly small hooves, a practice that has led to increased foot problems such as contracted heels and navicular disease. Florida veterinarians and farriers now are cautioning Quarter Horse breeders about such trends.

But the breed is due all the recognition it is receiving. These equines have an easy-going temperament, they are sturdy and unusually versatile, and intelligence tests given different breeds reveal the Quarter Horse is a quick learner and one of the smartest in the horse classroom.

Standardbred

The American Standardbred, or trotting horse, is yet another breed whose owners boast that it is the "all-American breed." In fact, Charles Marvin's 1890 volume, *Training the Trotting Horse*, maintains the trotting horse is the "ideal horse of business and pleasure" and should be designated as the U. S. "national horse."

Standardbred horses originated in Colonial America. It is generally thought that the grey English Thoroughbred, Messenger, imported to Philadelphia in 1788, is the originator of American Trotting Horses. It is from Messenger that the great trotting blood derived, a line that includes Mambrino, Hambletonian, the Clay family of trotters, and the Black Hawks (the Morgan line of trotters that included Ethan Allen).

The above Standardbred families all are trotters, that is, they move their diagonal sets of legs forward at the same time, left front leg with right hind leg, etc. Another group of Standardbreds perform a pacing gait; they move the lateral legs together, the left front and left rear legs moving forward at the same time. Famous pacing families include the Blue Bulls, the Pilots, the Hiatogas and the Copperbottoms.

Standardbred

Although trotting horse registration began in 1871, the breed called "Standardbred" technically began in 1879. That year the National Association of Trotting Horse Breeders required horses meet a performance "standard" in order to be registered. The standard was trotting one mile within two and a half minutes or less. Also eligible for registry were "parent or progeny of a standard animal." Standardbreds are usually bay, brown, or chestnut-colored. They are lightweight, attractively formed horses that average 15 hands tall and have generally kind dispositions. Inasmuch as most Standardbreds are bred and trained for racing, they can be difficult for backyard owners to retrain. For example, racing trainers discourage Standardbreds from loping or cantering; as a result, it is nearly impossible to later convince the ex-racer to canter. This factor has limited use of Standardbreds for showing. Outside racing circles, the breed is used most often for pleasure riding and by the Amish to pull their carriages.

Tennessee Walking Horse

Tennessee Walking Horses are mild-mannered, well proportioned and generally stand 15 hands to 16 hands tall. The breed traces back to a mid-1800s interbreeding of pacing horses, American Saddlebreds, Morgans, Standardbreds and Thoroughbreds. A single Standardbred stallion named Black (or Roan) Allan may be the breed's most famous forefather. It was not until 1935, however, that the breed's registry was established. And in 1950, the United States Department of Agriculture recognized the Walking Horse as a distinct breed.

The Walking Horse's way of traveling is somewhat similar to the movement of the American Saddlebred. In fact, the Walker, like the Saddlebred, was developed to provide the plantation owner with a comfortable, smooth ride. As such, the breed's best known gait, the running walk, clocks in at about 15 m.p.h.

Usually huskier and more athletic in appearance than a Saddlebred, the Walking Horse is fairly limited in shows to Walking Horse classes. An effort is under way to more widely display the breed's versatility, so that some Florida Walking Horse shows now feature trail and reining classes. Moreover, Walking Horse owners now are appearing more regularly on the trails and in long-distance endurance competitions.

Tennessee Walking Horse

At one time, show-quality Walking Horses were confined to those animals who could do the "big lick" — a far-reaching, high-stepping running walk. Unfortunately, the practice of "soring" a Walker's feet to achieve that degree of animation was widespread. Soring is the adding of caustic substances to the hooves to make the horse lift its feet higher and the horse appear livelier. The practice is now illegal. Moreover, the plantation style of riding Walking Horses — without built up shoes or artificial substances or devices — is more popular with show ring exhibitors than ever. Tennessee Walking Horses are intelligent, kind-tempered pleasure mounts that, despite their sturdy build and height, make good family horses, even for children.

Thoroughbred

Perhaps no horse breed has done more to put Florida on the world equine map than the Thoroughbred. The Sunshine State now ranks third in the nation in production of Thoroughbred racehorse foals, behind only Kentucky and California. In fact, nearly 600 Thoroughbred farms are situated in the state — each averaging nearly 175 acres. And when economics are computed, the final impact of the Thoroughbred industry in Florida tallies a whopping $1 billion. To examine the Thoroughbred, however, the clock must be turned back more than 250 years, to 1730, and the arrival in this country of Bulle Rock, generally accepted as the first English Thoroughbred in the United States. Records of Bulle Rock's siring successes are vague, but he inspired growing interest in importing English Thoroughbreds. Such newcomers were regularly crossbred with available Colonial horses that varied in quality from American Indian ponies to small mixed breeds of Irish, Scottish and English heritage. Whenever possible, the pure blooded imports were mated with other English Thoroughbreds. As a result, nearly all United States Thoroughbreds trace back directly to three 18th century English Thoroughbreds: Herod, Matchem and Eclipse. For example, Secretariat, Seattle Slew and Affirmed come from Eclipse lineage; the great Man O' War was a descendant of Matchem. Florida's relatively recent Thoroughbred industry must thank Eclipse for his genetic contributions to the state's first Kentucky Derby winner, Needles. When Needles was foaled, in 1953, only a dozen Thoroughbred horse farms struggled along in Florida, mostly in the Ocala area. His poor health as a foal necessitated many injections, which earned him his name. Nevertheless, he won the 1956 Kentucky Derby and the Belmont Stakes. Needles died in 1984 at age 31. Following Needles' victories, Florida's Thoroughbred industry bloomed. Today there are more than 500 Thoroughbred farms in the Ocala area. Those farms have produced such racing greats as Carry Back, Foolish Pleasure, Affirmed, Dr. Fager and Conquistador Cielo. As a breed, the Thoroughbred is extremely versatile and may be used in virtually all equestrian competitions except those reserved for special breeds such as Arabs and Morgans or for special riding styles such as Saddle Seat.

The best pleasure mount is often a Thoroughbred who has never been trained for the racetrack, or a horse that is part Thoroughbred such as a Thoroughbred-Quarter Horse mix. The breed comes in every solid color and stands 15 to 17 hands.

ESTIMATED BREED POPULATIONS

Florida's only equine census was made in 1971 and was based on the number of horses that received the Venezuelan Equine Encephalitis (VEE) vaccination. Subsequent horse population figures were based on those numbers plus an estimated growth rate of five percent per year.

Obviously, the state's current estimated total figures are skewed (most probably to the low side) for three reasons: 1) Not every Florida horse could have been vaccinated (and thus was not counted) in the 1971 VEE program/census. In fact, VEE is not routinely administered in Florida but was given in 1971 because of an unusual VEE outbreak. A more reliable census might have resulted if the count had been taken with administration of a more standard type of vaccination. 2) Not every breed grows at the same rate. Quarter Horses, Arabian horses and Thoroughbreds have grown much faster than five percent. And, despite breed registration figures, not all purebred animals and their offspring are registered because of lost papers or because they don't meet a breed's physical standards. 3) Growth of such breeds as Buckskins, Morgans, draft breeds and foreign breeds such as Peruvian horses and Trekehners are not reflected in previous charts. These breeds are listed under "other breeds." Moreover, some breeds were virtually unknown in Florida in 1971; therefore, subsequent figures have been blatant estimates.

The following chart, based on figures from the University of Florida's Institute of Food and Agricultural Sciences (IFAS), is the most reliable estimate of how many horses live in Florida. Figures, in addition to being estimates, do not reflect seasonal fluctuations in equine population (i.e., most foals are born January - April).

Breed	1981	1985	Value
American Saddlebred	1,700	*	$3,400,000
Appaloosa	4,079	4,590	8,263,710
Arabian	5,076	6,653	29,938,500
Paint	1,307	*	2,614,000
Palomino	2,750	*	4,125,000
Paso Fino	1,200	1,212	2,181,600
Quarter Horse	32,613	41,173	102,932,500
Standardbred	3,700	4,004	60,060,000
Tennessee Walking Horse	3,500	*	7,000,000
Thoroughbred **	20,020	20,500	1 billion+
Other breeds	3,500	*	3,500,000
Ponies of all breeds	15,341	*	1,917,625
Grades	100,000	105,302	63,181,200
TOTALS	194,786	216,032	$1,289,114,135

*Annual growth of less than one percent.

**Approximate figures from Florida Thoroughbred Breeders Association.

Tips and Quotes!

"The best color is fat." Old horse saying.

FLORIDA'S MAJOR COLOR BREEDS

To the average non-owner or non-enthusiast, a horse by any color is still a horse. And those colors usually are white, brown, black and grey.

Any avid horse fan will say, however, that those basics are divided and subdivided into a kaleidoscope of wonderful hues. The browns range from blackish to reddish, the blacks look bluish or auburn, the greys come in pinkish shades, gun-metals and nearly pure white. But the arrangement of those colors into patterns or other distinctive markings, as well as certain solid colors such as Albinos, Buckskins or Palominos, has resulted in cults of horse fans who get chills at the merest mention of their favorite color breed.

Color breeds often are considered as those breeds whose markings or colors may be more prized than their conformation. As such, then, a spotted Appaloosa or a satiny Palomino can resemble a sturdy Quarter Horse or a greyhound-sleek racing Thoroughbred. Other color breds, however, have strict conformation standards as well as color requirements. Regardless, one fact is certain: Color breeds are easy to identify. They are truly the horse world's horses of a different color.

Albino

Although less often seen in Florida, a few American Albinos do reside in the state. Some breeders and veterinarians believe Florida's sub-tropical sun makes life difficult for sensitive-skinned horses. This is unfortunate, for every horse person has at some time pictured being aboard a white charger. Or hi-ho-ing away on a Silver.

True albinos — with no hair, skin or eye pigment — do not occur in horses. What's more, two horses with dominant genes for pure white (homozygous) coloring produce an evidently lethal gene. So absolutely pure white animals die. The survivors who carry on the American Albino breed, then, are genetically impure and no true breeding strain has been developed.

The breed traces back in this country to the early 1900s. A single stallion named Old King had a knack for producing pure white progency when he was bred to solid-colored mares. Old King is said to have been an Arab-Morgan cross.

In 1937, the American Albino Horse Club was founded in Naper, Nebraska, and breeding records were part of the club's duties. Up to 1949, only genetically dominant white horses were recorded in the registry, although sub-sections of the registry permitted cream-colored animals sometimes called American Cremes. In 1970, the American Albino Association opted to separate the whites from the creams.

The American Albino may be found in any breed, including lightweight draft horses.

Appaloosa

Numbers of incredibly popular Appaloosas are soaring in Florida. At least one Appaloosa stands in virtually every large Florida horse barn.

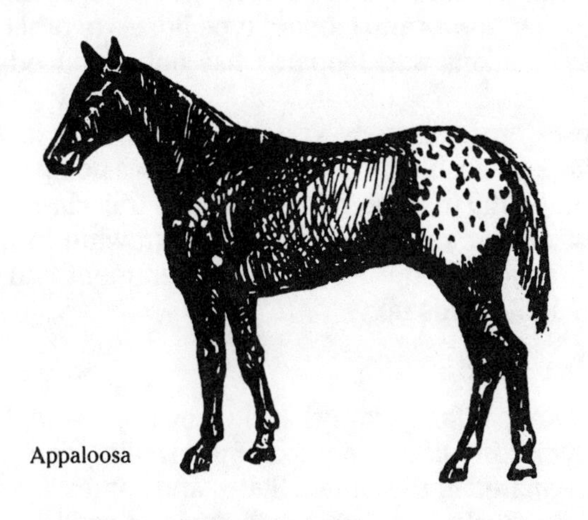

Appaloosa

Appaloosa horses are first recorded in North America as Indian horses especially prized by the Nez Perce Indians. The native Americans, in fact, organized the first Appaloosa breeding program. Worldwide, Appaloosa-type horses appear in Chinese paintings that date back to the Tang Dynasty (618-907 AD). Appaloosas nearly became extinct in this hemisphere after the Nez Perce War of 1877. The Indians lost their battles with the U. S. Army and their horses were sold off. Fortunately, the Indian horses also found favor among the new North Americans and breeding programs saved the breed.

Four general characteristics are the hallmark of the Appaloosa breed. Most obvious is the coat pattern that may vary a great deal from spotting all over the body, to animals that are mostly solid with a spotted "blanket" over the hindquarters, to slight specks on a solid horse. The so-called leopard Appaloosa is colored similarly to the Dalmatian dog, with a white background and rather evenly-distributed dark brown or black spots.

A second breed characteristic is mottled skin, that is, the horse's skin has irregular spots of black and white, most noticeable around the nostrils. Third on the list of requirements is hooves that are striped vertically with black and white pigment. A fourth Appaloosa characteristic is an eye that is similar to a human's with a white sclera around the eye's brownish-black iris.

Breeding two Appaloosas does not always produce predictable results. For this reason, breeders recommend breeding blanket horses to blanket horses and leopard type horses to other leopards. Crossing the blankets and leopards has not produced favorable results.

Appaloosas are extremely versatile horses. Any local Florida show will have a sizeable turnout of Appaloosas going over hunter fences or competing in Western pleasure and trail classes (or both). And enthusiasm for racing Appaloosas is growing in popularity. The breed is renowned for its docile temperament and ranges in height from 14.3 hands up.

American Buckskin

Any Roy Rogers fan recalls that the cowboy's wife, Dale, rode a pretty Buckskin horse named Buttermilk. The American Buckskin originated in the United States and comes in a variety of colors including yellow buckskin, red dun and grullo. Genetically speaking, the Buckskin coloration is a dun. And dun is a dilution factor that "waters down" other solid coat colors such as black or brown. The result is attractively muted yellowish tans, reddish browns or smokey ("mousey") dun. True Buckskins have black skin under their hair, and have a darker brown stripe down their backs, dark lower legs and nearly black manes and tails.

Buckskins also are very versatile — any breed of horse can be registered as a Buckskin as long as no Appaloosas, draft breeds or pintos are in the geneology. And Buckskin afficionados are quick to point out they do take conformation as well as color into consideration.

This breed performs every type of equestrian activity, but in Florida is a special favorite of Western-style riders.

Paints and Pintos

Despite the fact that dictionaries make Paint horses and Pinto horses near synonyms, the owners of both these breeds (as well as their registries) are quick to point out the distinctions. In fact, they wince at how frequently horse-ignorant folks get the two breeds mixed up.

Perhaps the spotty record can be set straight in alphabetical order.

At least one Paint horse reportedly arrived in North America with the Spanish conquistadors. Indians adopted the prettily marked animals and later the cowboy favored the breed's flashy markings. As cars took over the transportation chores after the turn of the century, numbers of Paint horses dwindled. Fortunately, some individuals actively bred the horses. In 1962, the American Paint Stock Horse Association was founded in Texas. Three years later the American Paint Stock Horse Association merged with the Paint Quarter Horse Association to form the American Paint Horse Association (APHA). By 1979, 611 APHA-approved shows were held with more than 125,000 entries.

The APHA is devoted to Quarter Horse (stock type) horses. Registered Paints, Thoroughbreds and Quarter Horses are the acceptable breeding stock. In fact, Paint horses must present documented parental evidence of any two of the three aforementioned breeds. Paint horses emphasize conformation over color. However, the Paint horse must be white with any other color and must have a recognizable paint pattern.

The Pinto horse breed also claims Spanish ancestry. In appearance (stock type), Pinto horses may resemble Paints but the Pinto Horse Association of America is more lenient in its admissions. Four types of Pinto horses are eligible: the previously mentioned stock-type spotted horse with Quarter Horse or Paint breeding; the pleasure type horse with Arabian breeding; the hunter-type Pinto with Thoroughbred breeding; the Saddle-type animal that has Saddlebred breeding.

Both Paints and Pintos come in two pattern models, Tobiano and Overo.

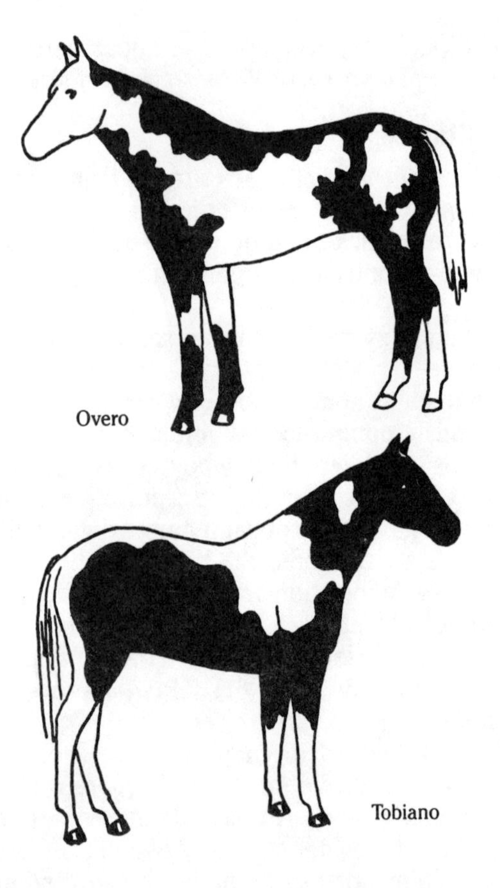

The Tobiano has "smoother" looking spots on the head, chest, flanks and in the tail. The legs are almost always white. Overo markings are lacy or jagged-looking, mostly in the middle of the body. The legs usually have more color than white and the white rarely crosses the back.

Both breeds are extremely useful in a wide range of activities. Florida's Paint and Pinto population in Florida is on the increase.

Palomino

All sorts of images pop into the mind when conjuring up the golden Palomino. Perhaps no other color of horse is as readily identified by the general public as is the Palomino, due in great part, no doubt, to Roy Rogers' famous Trigger.

The word Palomino is Spanish for "like a dove," and the breed is thought to have arrived in the New World with Cortez. The Palomino's distinctive golden color with silvery mane and tail occurs in several breeds, although most rarely in Thoroughbreds or Arabian animals. Half-Arabs and half-Thoroughbreds do sometimes come up with Palomino progeny. Color requirements of the Palomino Breeders Association mandate no dark or light hair on the back. There also cannot be any stripes. White on the face and lower legs is permitted. And Palomino horses may have hazel eyes. Most Palominos are stock horse types, although Saddlebred type parade horses also are popular.

Breeding to get a Palomino can be a frustrating experience. The color breeds true only about half the time because of a genetic dilution factor. Thus mating two Palominos four times predictably produces one Albino foal, one chestnut, and two Palominos. To obtain a Palomino foal, geneticists recommend breeding an Albino with a chestnut. Palominos, ideally 15.2 hands, are most popular for Western-type classes but are competitive in the hunter ring, especially at the local level in Florida.

FLORIDA PONIES

The British have a charming habit of starting their children out learning to ride ponies instead of full-grown horses. Of course, when children reach their teens they usually outgrow their ponies. And when that time comes, it provides a good opportunity for parents and teens to decide whether they should purchase a horse and continue riding.

A pony "starter" is a good idea for many reasons. Ponies are easier kept, demanding less food and less space than a horse. In addition, ponies are usually a little hardier than horses and are sick less often. Perhaps the most attractive aspect of ponies is their cost. They nearly always carry a lower price tag unless they are veteran show ring champs.

Technically, a pony is an equine that stands 14.2 hands or smaller, measured at the withers. However, some diminuative horses, especially Arabians but also Morgans and some Quarter Horses, may measure 14.2 hands or a little less, but are not officially classed as pony breeds. In large shows, especially English hunt seat and those shows with pony classes, the pony must possess a card that proves the animal has been measured by a steward or judge and is indeed 14.2 hands or less.

Many people believe that a pony is a young horse — or is synonymous with a horse. This misconception may have come from television and books about the West. Cowboys had their ponies. While there is some debate as to whether these cowponies were truly ponies or not, they definitely were wiry and hardy and usually of no recognizable breed.

In fact, height in horses may be a by-product of domestication. For when left in the wild, equines tend to stay ruggedly compact — 12 to 15 hands — probably to increase their efficiency.

True ponies, especially purebreds, have distinctive conformation traits and may tend toward draft types or saddle types.

Shetland Pony

Best known of all pony breeds is probably the Shetland. It originated in the Shetland and Orkney Islands, probably in about the 6th century. British versions of the Shetland pony are about 42 inches at the withers. The U. S. version is slightly taller, about 46 inches, with a more refined appearance.

Shetland Pony

British drive carts behind Shetland ponies, but do not usually recommend the breed as children's mounts. One reason is that Shetland ponies are too small to be trained under saddle by adults, so they may be too difficult for children to handle. Another drawback is that such a small pony will be outgrown too quickly.

More suitable ponies for children have a little more size and have been trained by adults. The following ponies are growing more prevalent in Florida:

Connemara Pony

Named for its western Ireland homeland, the Connemara is hearty, well dispositioned and stands about 13 to 14.2 hands tall. The most common color is grey, from flecked (flea-bit) to lush gun-metal. The breed has delicate heads, sloping shoulders and compact bodies that combine with gentle temperaments and intelligence to make a good child's mount.

Breed standards for Connemaras were updated in 1980. The once rather "heavy-and-hairy" type ponies have been refined through some cross-breeding with Thoroughbreds. The ponies make suitable mounts for small adults, too.

Hackney Pony

These most attractive ponies are not often seen under saddle in pony classes. More often, they are put to cart and competed by adults. Hackneys range in size from 11.2 to 14.2 hands tall. For showing, the Hackney pony has a high-stepping action.

Pony of the Americas

These "mini-Appaloosa" type ponies are becoming more popular in Florida. They range in height from 11.2 hands to 13.2 hands. Pony of the Americas usually are shown Western-style, but make good pony hunters. They generally have quiet dispositions.

Welsh Mountain Pony

These ponies rank among the most popular in Britain and in the United States. Bred originally in the mountains of Wales, the breed developed hardiness and sure-footedness. Welsh Mountain ponies stand about 12 hands high and are extremely intelligent, a fact that occasionally makes them overly challenging for children. But they are willing to learn and lend themselves well to pony hunter courses and to driving.

A slightly larger pony, standing 13.2 hands, is the Welsh Pony. Many pony breeders belive the Welsh pony is a more-or-less "improved" version of the Mountain Pony. It also is very popular in the United Kingdom and growing more popular in Florida. These ponies are considered "cob" type ponies.

Pony purchasers should follow the same guidelines as for buying any equine. A veterinary checkup is always wise, even if the pony has a low price tag. Children get attached to their ponies and it can be a traumatic experience if a new pony grows sick and dies.

Ponies are some of the equine world's most unique and lovable "short stuff." Many riders, even those now competing at the Olympic level, still have their old pony friend back home in the barn. That could be because ponies are usually very intelligent and frequently brim with personality. A good show pony often puts full-sized horses to shame with flawless performances.

Riding is knowledge a person keeps for a lifetime so starting with the right animal is an important first step. And there's no better friend or riding teacher than a good pony.

IMPORTED BREEDS FOUND IN FLORIDA

The past decade has seen a startling increase in numbers of foreign bred horses. Such breeds as Trakehners, Swedish or Dutch Warmbloods, Holsteiners, Paso Finos and Peruvian Pasos now are becoming familiar to most Florida horse fans. Many imported breeds, such as the Pasos, have their own large shows in the state, and are treating spectators to spectacular and beautiful new costumes, music and styles of riding. Ever more popular among hunter, jumper, dressage and eventing riders is interbreeding of Thoroughbreds with such sturdy imports as Trakehners, Warmbloods and Holsteiners.

Just as with so-called all American breeds, the imported stock have rich and varied histories and doubtlessly will impart their own special contribution to Florida's horse world.

European Breeds

A variety of European horses are just beginning to show up in Florida horse barns. With names that many Floridians cannot even pronounce, these breeds have long been popular in Europe for sporting and pleasure riding.

Many European horses are classified as "warm bloods" because they are a blend of "cold-blooded" draft working breeds and "hot-blooded" Arabian or Thoroughbred ancestry. European registries generally require prospective registrants to pass conformation, performance and disposition tests, particularly if used as breeding stock.

Most familiar to Florida horse people is probably the Trakehner (pronounced tra-CAIN-er), a German breed founded in 1732 on the farm of Trakehnen in East Prussia. The Trakehner breed is an exception among European-bred animals because its primary use has always been as a riding horse.

That the breed survives today is a story worth relating. During World War II, many Trakehners horses — some sources say 90 percent — were killed by Allied bombing raids. To save the remaining stock, locals led the Trakehners on a 600-mile journey over ice and snow across Prussia and Pomerania to the relative safety of northwestern Germany. By the time the horses arrived at their new home, many were fleshless frames with feet wrapped in frozen burlap bags, say eyewitnesses, and carried open wounds from bombs and gunfire.

Trakehners are elegant, sturdy, good natured and large — usually 16 hands or taller. Their most publicized uses are as grand prix jumping horses and as dressage mounts.

Holsteiners and Hanoverians also are finding Florida homes. Both these German breeds often were used as carriage or artillery horses.

Holsteiners are said to date back to the 14th century and were bred originally to be warhorses. Subsequent breeding with European Thoroughbreds refined the breed and resulted in a popular undersaddle animal. In recent years, the Thoroughbred again has been used to further refine the Holsteiner, but it remains a laid-back, sturdy and large (16.2 hands and up) mount. Like the Trakehner, the Holsteiner is becoming a favorite jumper and dressage horse.

Hanoverians won the favor of the horse-crazy Britains in the 18th century when King George, Elector of Hanover, reigned. Although the Hanoverians were used chiefly as carriage horses, breeders soon found that crossing British stock with the sturdy

German breed produced a highly desirable 16-hand, or bigger, riding horse. Hanoverians often are seen on the jumper courses or in the dressage arena.

Hanoverian

Other, less-often-seen, European mounts include Swedish and Dutch Warmbloods. As with all European horses, the name usually indicates where the breed originated.

Swedish Warmbloods are the result of interbreeding Trakehners, Hanoverians and Thoroughbreds with native Swedish stock. Dutch Warmbloods are a cross between a variety of European breeds with Dutch Gelderlanders. Both breeds also are used for pleasure riding, jumping fences and dressage.

Lipizzaners can be found in Florida, primarily in the dressage arena. The breed's famed white stallions still are used in Vienna's Spanish Riding School and perform the difficult "airs above the ground," which are battle maneuvers. The breed was founded in 1580 and was used originally as war chargers. It is smaller than most other popular European imports, standing 14.2-15 hands high.

Like the Trakehners, Lipizzaners came close to oblivion during World War II and were evacuated by Gen. George Patton. The breed's story was told in the Walt Disney film, *The Miracle of the White Stallions.*

Latin American Breeds

Two Latin American breeds have particularly appealed to Floridians during recent years. They are Paso Fino and the Peruvian Paso. Both breeds have a separate registry although many references to them meld the two breeds. Paso Fino literature says Columbus took Spanish Jennet horses (known for their smooth saddle gait) to Santo Domingo in 1493 where they became remounts for the Spanish conquistadors. Slowly, the breed spread throughout the Caribbean — to Puerto Rico in 1509, to Cuba in 1511, to Panama in 1514. Out of the Spanish Jennet line came the smooth-riding Paso Fino with its subtle differences in the unique "Paso" gait. Puerto Rico, Cuba and the Dominican Republic prefer the "sobre paso," a relaxed, long-reined gait or a very speedy "paso largo." Plantation owners, however, like the very slowest form of the gait, "paso fino."

Paso Fino

Peruvian Pasos also came from Spain — but went directly to Peru for use by that nation's noblemen and politicians. Both breeds have what is to North Americans a unique way of traveling. At first, they appear to be pacing — with the two legs on the left side of the body moving forward at the same time — then the right pair of legs moving forward together. But the legs are set down in a four-beat gait in a regular rhythm, like taps on a drum. As a result, the rider is barely jarred in the saddle as the horse glides along smoothly.

But there the similarities in movement end. The Paso Fino takes small, fine steps. Forward movement is minimal, though the horse may be taking hundreds of steps in a 25-foot-long path. This "fino" gait is highly prized and is performed in shows on a "fino board" — a wooden walkway that amplifies the castinetlike footfalls. The Peruvian horse covers more ground with each step. Peruvian animals also throw their front feet out to each side more than the Finos do.

Some conformation differences exist between the two breeds — the Peruvian horse appears slightly larger with perhaps a little more angle to the shoulder and hindquarters. Both breeds have magnificently thick and curving necks, long flowing manes and tails. Although both are smaller than many North American breeds, about 14.2 to 15 hands, they are amazingly sturdy with great stamina. They are basically kind and even-tempered.

Florida's climate, pasture, and receptive horse lovers bode these Latin American breeds a good future.

DRAFT BREEDS

Newcomers to Florida's horse world rarely think of the state as a haven for the heavy draft breeds. But populations of a variety of draft breeds is on the upswing in Florida. Many light horse owners enjoy having a draft horse around, while some north Florida farmers still use the big equines to pull farm implements. Other Florida owners exhibit their drafts either at halter or in driving competitions.

Even Florida tourist attractions are home to draft breeds. Walt Disney World uses several breeds, mostly Belgians, to pull the trams down Main Street, and has one of the nation's finest Percheron hitches that is used for parades and for Draft Horse Hitch contests. Likewise, Tampa's Busch Gardens is frequently homebase for the Budweiser Clydesdales, probably the most famous draft horses in the world.

Generally, draft horses developed in Europe where plentiful fodder proffered equine behemoths. As large animals, draft breeds were valued for their brawn and not their speed, although onlookers are occasionally stunned to see the rapid pace and agility the draft horse can generate if necessary.

During the Middle Ages, the draft breeds became the mount of choice for armoured knights. Indeed, few lighter breeds could have supported a knight, for in full armour he weighed about 300 pounds. In jousting matches, the heavier horse lent better support.

Drafts are usually considered "cold bloods" in the equine world, but the Percheron is believed to have a little hot blood in its ancestry. This is thought to have occurred when the Moors riding Arabian horses overran the Percheron's native France. It is thought some Moorish horses were crossed with the Percheron.

The word "draft" comes from pulling heavy loads — the work most commonly assigned the heavy horse breeds. In Florida, passersby can spy draft teams working a few rural farms. Draft horses are capable of moving 7,000 board feet of lumber in one day. Of course, logging tractors can move four times that much lumber, but cannot navigate around thick woodlands as well as the horse. Plus, the horses are certainly more aesthetically pleasing.

Following are some of Florida's most popular draft breeds.

Belgian

As the name indicates, the Belgian draft horse originated in that country. These horses are the most compact of the draft breeds, averaging about 16 hands high and weighing one ton.

Belgian

Farmers who still use draft animals often favor Belgians because they are shorter, closer to the ground and with a lower center of gravity that makes them able to pull heavier loads than taller animals. Belgians are commonly chestnut, dun and red roan colors. They mature early, are docile and easy to handle.

Clydesdale

The most recognizable draft breed, probably due to television advertising, the Clydesdale is the show horse of the heavy set. It originated in Scotland, along the Clyde River, and was first imported to the United States in the 1800s. Despite its size, usually about one ton and 16.2 to 17 hands tall, the Clydesdale moves in an animated style. Most usual colors are chestnut or roan with striking white markings. Black or grey Clydesdales are occasionally seen. Because they are so showy, Clydesdales are a popular driving horse for exhibition or for parades. The breed often pulls six-horse hitches.

The Clydesdale is less favored by farmers as a work animal because of its height and because its fluffy white feet require frequent grooming and upkeep. Clydesdales are alert, but gentle and intelligent. Increasingly, the breed is being crossed with lightweight horses to improve their stamina and sturdiness. Progeny of such crosses are used in eventing and show jumping.

Percheron

Percherons are another fluffy-footed draft breed. They come from the French district of La Perche. At one time the breed boasted great popularity in the United States, especially in the nation's farm belt. Virtually all Percherons are black or grey. The breed's movement is less animated than the Clydesdale's, but the Percheron often has a more refined-looking head, a physical trait thought to harken back to its Arabian ancestry. The breed is easy-going, long-lived and stands 16 to 17 hands tall. It also tips the scale at about one ton.

Prospective purchasers of any purebred animal should be aware that breed registries occasionally change their admission requirements. Newcomers should always check with local breeders or should write a horse breed's national headquarters for the most specific and up-to-date information.

Chapter Four

In Health and Sickness

A little neglect may breed mischief; for want of a nail the shoe was lost; for want of a shoe the horse was lost; and for want of a horse the rider was lost.

— Benjamin Franklin

Horse-owning Floridians are lucky. A most expensive facet of keeping any animal — especially one the size of a horse — is housing it. Fortunately, such costs are minimal in the balmy Sunshine State.

Basically, horses don't have to be stalled. Many horses are kept full-time in pasture, a practice endorsed by some veterinarians who say these all-outdoors equines are probably the healthiest and the happiest. Pastured horses not only exercise more, but they can eat whenever they want of what they enjoy most, grass.

Moreover, cooping up a 1,000-pound animal in 144 square feet of stall all day can contribute to the equine equivalent of cabin fever. Nervous-tempered horses get edgy, even hard to handle. On the other hand, especially fractious horses have turned into veritable pussycats when turned out to pasture for long, leisurely grazings.

Horses that are stalled much of the time also tend to form bad habits. The mildest-mannered dobbins are known to grow bored and take up chewing wood, pacing, digging, kicking and weaving.

Such habits are hard to cure. Additionally, they devalue the animal in the eyes of a prospective buyer.

STABLING

Regardless of the consequences, some owners still prefer to stable their horses. These owners may believe they do not have good enough pasture grasses to support their steed. Or, if the horse is exhibited regularly, the owner does not want the animal's coat to sun-fade. True, stalled horses do stay cleaner and more free from scrapes and injuries, especially if turning the animal outside would put it with other horses who bite or kick.

And there are times that stabling — or at least sheltering the horse — is a necessity. Gaited horses with built-up feet, for example, may injure themselves in a pasture because they are not as sure-footed as flat shod horses. Sick horses, or those animals that are allergy prone, should be stabled. Understandably, too, most owners prefer to keep a mare whose foaling time is near, safely secured.

In Florida, a primary reason to ensure a horse has at least minimal shelter is the many, often violent, thunderstorms — and the lightning that often accompanies such downpours. The Sunshine State can get downright un-shiney sometimes, with an average rainfall of between 50 and 65 inches — an estimated daily total statewide of 148 billion gallons. In summer, there is a 50-50 chance a locale will be rained upon each day. Frequently accompanying summer rain is lightning. And Florida lightning is no trifle, a single bolt being capable of delivering a 6,000- to 10,000-amps punch. Studies reveal the state is the thunderstorm and lightning capital of the nation, recording the nation's highest number of human deaths due to lightning — 227 from 1959 to 1980, reports the National Oceanic and Atmospheric Administration. Pastured horses in Florida are just as vulnerable to being killed by lightning as humans in unprotected areas. Often the horses have taken refuge under a tree that acts as a natural lightning rod, delivering the strike directly to the animals. Other horses become equine lightning rods when they are the tallest objects in a field. Thunderstorms also yield the additional dangers of hail and tornadoes. Then there is the other extreme — when it's not raining the blistering sun can overwhelm horses, even in winter. Shelters, therefore, afford the horse a cool resting place.

Most infamous, and publicized, are the state's hurricanes. It is a secure feeling for a owner to know ol' equus is high and dry waiting out such a storm in a sturdy stall. But an owner cannot become complacent, for once a hurricane forms it poses not one, but three, threats to humans and horses, alike. Winds gusting 75 miles or higher can uproot trees and lift barn roofs. Near the Gulf or the Atlantic Ocean, wave action can tear structures apart and flood out what is left standing. And third, heavy rains add to flood dangers.

Horse owners whose horses are stabled in low-lying areas, particularly near flood-prone water, should, if there is time, try to move their horses to higher ground. Usually this is not possible because all the alternative shelters are in the same predicament — even if there is time to trailer a horse from its homebase.

Statistics reveal chances are greatest of being hit by a Florida hurricane in the Miami and Key West areas — one chance in seven. Horseowners near other coastal cities have a lesser chance of being in a hurricane's path: Palm Beach, 1 in 10; Tampa Bay area, 1 in 20; Daytona Beach, 1 in 30; Jacksonville, 1 in 50. Inland residents are safest, of course, because damaging hurricane winds often diminish over land and swollen tides are usually not a factor. Clearly, most of Florida's inclement weather comes in the form of moisture. Cold weather, a major weather phenomenon confronting horse owners in the north, rarely poses a problem in Florida except in the state's northernmost areas. Harmless frost does occur, albeit infrequently, throughout the peninsula; as for hard freezes, they seldom cause much damage other than occasionally popping a water pipe. If owners choose to blanket their horses, for instance, only lightweight rugs are necessary.

THE BACKYARD HORSE

Horse owners who are fortunate enough to have enough land for a horse or two are often termed "backyard" owners. Enough land for horses pastured all the time in such Florida backyards, experts say, is approximately two acres per horse, at a minimum.

That such acreage should be fenced goes without saying. But what does need saying is: *not with barbed wire.* Fencing that is available in other areas of the nation, with the exception of stone fence rows, is available in Florida. That includes wood stock fencing, board fencing, chain link fencing, high-tensile wire and the new plastic fencing that resembles wood.

Barbed wire is all right for cattle, but not horses. Yet one would be hard put to find a Florida veterinarian that has not spent time stitching up some horse's barbed wire wounds.

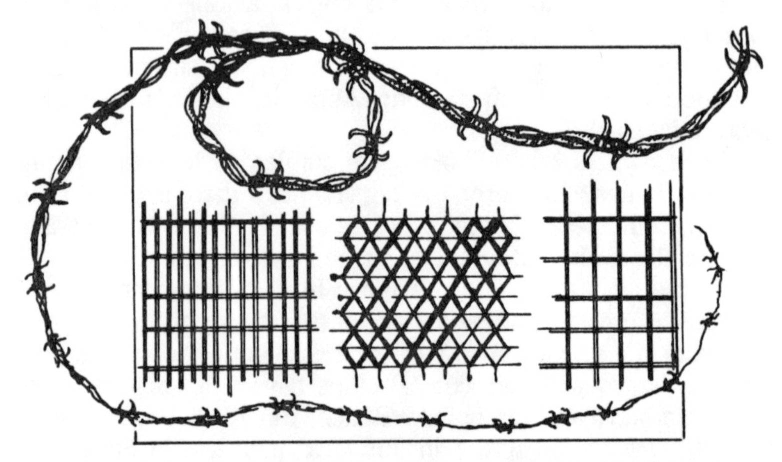

Fencing materials (left to right): Ordinary wire mesh stock fencing; diamond wire; close-woven stock wire. Around the edge of the picture is undesirable barbed wire.

No knowledgeable horse owner would string barbed wire around a pasture of horses, but such wire often is seen surrounding grazing horses in Florida. Usually, the barbed wire is there because it was already on the property when the horse owner purchased it. A false economy is then employed by some horse owners — rather than replacing the wire they tie little white cloth strips to it, believing the horse now can see the wire and thus avoid it. Not only do those little white cloths look tacky and betray a horse owner's cheap fix, but a truly panicked horse on the run may still run into the dangerous obstacle. And once that occurs, and the veterinarian's bill arrives after the stitches are taken, barbed wire no longer looks so economical.

Wire stock fencing, like barbed wire, is one of the lower priced fencing methods. But unless it is expertly installed, it will sag in a year or so and must be restretched. The procedure is not difficult but it is time consuming if pastures are large. Several types of wire stock fencing are available, their differences usually being the size of the wire and the design of the wire squares. Be wary of wire

Paint can be easily added to the top of a wire fence with a roller; thus, the horse can see the fence better without the owner resorting to messy-looking white rag strips tied to the fence.

squares so large that they can snare the small feet of a foal or pony.

The gauge of plain stock wire is usually too weak and its square design too open to make a good fence for horses. Better wire fencing is made with two-inch by four-inch squares. Or, better yet, is trimesh wire, sometimes called "diamond" wire. Its triangular weave gives extra strength and horses cannot get anything caught in it (except maybe their halters, which should not be left on the pastured horse).

All wire fencing must be fortified with a wood board along the top, or with electric wire. Otherwise, when a horse leans over the top fence line several times, the top wires are bent down and the horse can easily step over the fence. Board fencing, either three-board or four-board, is the most expensive type of fencing and requires some upkeep. Not only does the weather get to wood fencing but some horses chew it or split the boards with their kicking. Anti-chewing paint is on the market, as are replacement boards.

A partially finished wooden fence as viewed from inside the pasture. Stagger the nailing of each level of planking to provide extra stability. If one board comes loose, or one post loosens, some planks remain sturdily in place.

Another disadvantage to board fencing is that dogs and other small animals can get underneath the lower board and into the pasture. To keep out such unwanted visitors, mesh fence is attached to the lower board.

Extra boards, stains and treating do add to the cost of board fencing, yet most horse authorities (and real estate folks) agree that wood fencing costs are usually recouped if the property is put on the market because the board fencing is attractive and adds to property value.

Properly installed wooden — or wire — fencing should have posts placed on the outside of the pasture. If the horse decides to lean on the planks or wiring, the posts provide crucial support. Otherwise, if planks or wire run along the outside of posts, and a horse presses against it, the nails or staples will pop loose.

Chain link fence also is a good choice, especially for stallion or foal paddocks, but like board fencing, it too is expensive to purchase and have installed. Chain link is weatherproof, requires low maintenance and is adequately sturdy, although an itchy horse can eventually push and pooch such a fence out of place. Gaining popularity in Florida among horse owners is high-tensile electric fencing. With only a few strands of wire, it looks tidy, requires little upkeep and is fairly economical. A pulsating current of electricity

Tips and Quotes!

An old-time cockroach chaser: one cup of dry plaster of paris mixed with two cups of dry oatmeal and one teaspoon of sugar. Sprinkle the concoction into buggy hiding places — out of the reach of horses or barn pets.

---

maintains the fence's integrity as well as the horse's respect.

Fairly new on the market and in Florida is a plastic fencing that looks like three- or four-board wood fencing. It comes in white, brown and black colored plastic. Attractive, the fencing reportedly will not fade in the sun so it is an easy maintenance fencing choice. Additionally, the manufacturers have selected a tough, resilient type of plastic that will withstand a horse's chewing or kicking. Plastic fencing enhances property values, but is quite expensive to purchase and install.

Ideally, full-time pastured horses should be given some enclosed area, even if it is just a "run-in" shelter, to escape the weather perils mentioned earlier. Such shelters need not be fancy but should provide about 12 square feet of space for each horse. Four posts or poles that support a high (10 feet or more) sloping roof, enclosed on three sides, is adequate. The building should be sturdy with sides made of either concrete block or wood planking. Dirt floors are usual in such structures, and acceptable.

A fancier version of the basic run-in shed has its fourth side partially enclosed to provide a doorway. This opening then can be closed off with a gate or a stall guard in case the horse is being confined because of illness or is being administered to by a farrier or a veterinarian. Next step up, in size and cost, from a run-in shed is a horse barn. Dozens of layouts are available from commercial companies in Florida, and nationally, whose primary business is building horse barns. Similarly, building materials come in a variety of choices. Regardless of the exterior design and the materials used, stall size deserves major consideration. A full-size horse usually needs a minimum 12-square-foot abode, while a pony or small horse can be comfortable in 10-square-foot living quarters. No matter what the stall size, however, each one should offer a window and good ventilation. Wood is excellent stall building material for Florida. It is readily available in the state, or from nearby Georgia, and usually reasonably priced. In addition, it insulates, it is attractive and it can be assembled by any handy horse owner who chooses to do his or her own construction work.

Of course, wood stalls can be kicked apart or chewed up by devilish equines, but a good builder, taking such horse vices into account, can compensate for such possible destruction by strengthening stall construction. Ironically, plywood makes good stabling material because it does not shrink and is durable. Horses have a difficult time kicking it apart. It is also less expensive than wood planks. The problem is it looks like plywood — a factor that turns many horse owners off. All wood burns so fire prevention must be strictly enforced.

Another popular Florida stall material is concrete block. Block costs less than wood, but hiring a mason to build the barn can actually increase the per-foot cost. The primary drawback to concrete barns in Florida is that they hold dampness. Other owners object to concrete's hardness — when a horse kicks block, it can damage the hooves.

Yet other barns are made of steel or aluminum. Easily and quickly assembled, they also are usually less expensive than wood barns and are easy to keep clean. Primary objection to metal or "pre-fab" barns is their lack of ventilation, a significant factor in Florida. Fortunately, a former flaw — the "industrial park" appearance — is disappearing. More traditional styles of these steel or aluminum barns now are available from a variety of firms.

Count on a horse barn averaging about $1,500 per stall to build. Naturally, the more amenities, the higher the cost. Such luxuries as ceiling fans, automatic insects sprayers, sliding doors and individual stall lights all make a barn attractive and more convenient. Add other such goodies as a grooming area, hot water and plenty of electrical outlets and the per-stall cost can skyrocket.

Whether the horse owner has the finances to provide a horse mansion or not, the horse barn should meet certain basic requirements.

Building on a high and dry site is one. In Florida, that decision should come after selecting several possible sites and then inspecting them immediately after a heavy rain. Using as much fire-proof material as possible is another. Again, with Florida recording such high numbers of lightning storms, fire is a definite hazard. Consider installing lightning rods on the barn to distract bolts from the barn itself. Other requirements include: hay storage that is separate from the horse's living quarters; aisles that are roomy enough — 8 feet or more — to lead a horse past stallmates without getting nipped; stall doors should measure four to five feet, preferably five, which allows both the horse and its handler to enter a stall at the same time; doors should always open out, not into, a stall (and cover stallions' doors with chain link-type wire or grillwork to keep them from biting passing barnmates). Regarding ventilation, many Florida owners keep their horses cooler in summer by leaving one-inch gaps between wood plank siding. Also keeping down a barn's summer temperature are airy ceilings, at least 10 feet high, to help hot air rise above a horse's head. Of course, such stalls are cooler in winter, but in Florida that still means "mild." Fortunately, it is rarely necessary in the Sunshine State to completely close up a barn because of cold temperature. Horse owners new to Florida often marvel at how seldom their equines are sick in their new home; the openness the year round of Florida barns is the major reason.

One small but significant detail that some barn builders forget is that horses are social beasts. It is always best if barn mates can see each other. With that in mind, chain link fencing or heavy duty mesh between stalls, at head level, permits horses to socialize without fighting with each other.

As for stall flooring materials, Floridians can run the gamut. The basic, of course, is leaving the floor alone. But this is not the best choice, because urine dampness soon makes the stall filthy. Horses dig it up and soon it is one large, wet pothole that lends itself to thrush and flies. Good stall flooring should be dry and either completely level or slightly convex to promote moisture away from the center.

Clay is a common and popular stall floor foundation in the South. It drains well, is slightly resilient under the horse's feet and is hard enough to resist equine excavators. But with time and wetness, clay surfaces can become cement hard and difficult to dig out and replace. For barn owners in southern Florida there is also the cost factor; clay is not native to their part of the state but has to be trucked in from north Florida or Georgia. Another oft-used Florida flooring is limestone. Like clay, well tamped limestone drains well, is resilient, resists digging and is a natural stall sweetener.

Poured concrete is occasionally used by well-heeled Florida

horse owners. Some concrete stalls even include drains so that cleaning — even disinfecting — can be done easily. The less well-to-do should be aware that concrete does not drain too readily by itself, that cleaning must be done frequently or the foundation becomes slimy and holds a urine odor. Biggest drawback of concrete, however, is its hardness. Unless it is well padded with bedding, it becomes uncomfortable for the horse to stand on for long times. And a digging horse can wear down its feet on the hard surface.

As for stall bedding, it can be straw, wood chips or wood shavings, in fact, anything that is absorbent and easily replaced. Owners should select bedding as carefully as they buy feed because some horses chow down on their bedding. Whatever material is used for the horse's bedding, it must be dry, clean and as dust-free as possible. Straw should be well cured and not moldy. Wood chips or shavings also should be cured and neither teakwood nor walnut wood used; these two woods can be toxic if the horse eats its bedding accidentally or on purpose.

Ironically, bedding materials have become hard to find in recent years and, when located, may be expensive. For this reason, the Florida Agriculture Department is researching new stall bedding possibilities. Oat and wheat straw are being investigated, but holding the greatest promise is plain old paper. Experimental paper bedding has been made from shredded, uncirculated telephone books. So far, research reports are favorable with the tested paper bedding revealing it is more dust free and more absorbent than traditional straw and wood products. In fact, tested paper bedding is turning out to be twice as absorbent as straw and nearly five times as absorbent as wood shavings. Other plusses for paper is that it is cushioney and so far has not harmed the horses that ate their experimental beds.

Nothing is on the horizon, however, to lessen the stall chore of cleaning up a horse's droppings. It is something that a responsible owner does every day. If the horse is never turned out to pasture, the stall needs picking up even more often. Of course, urine-soaked foundations demand cleaning out whenever the stall starts getting smelly.

Not to be overlooked is a stall that is minimally and safely equipped for a horse's comfort. A water bucket and a grain bucket are the only absolutes, and it is not a good idea to use cheap metal buckets. They get dented and seem to sprout sharp edges. Smooth corner molded plastic buckets are best. Salt — regular white blocks and red mineral blocks — should be in the horse's stall at all times. Many owners like to install block holders near the grain bucket.

Good style rubber or plastic grain tubs have rounded edges that prevent horse injuries.

On the "may be necessary" stall list are rubber mats in front of the stall door to discourage diggers. Hanging horse "toys" — clorox bottles or plastic balls — from the stall ceiling entertain some horses, owners contending these items discourage some wood chewers or weavers from practicing their vices. Be prepared, however, for a horse, like a child, to tire of his playthings. A more controversial stall fixture is hay, specifically, where to place it in the stall. There are three choices: a hay rack, a hay net, the stall floor.

A hay rack is usually made from metal, in a grid, that fastens to the stall wall to hold one or two flakes of hay. Racks must be attached a little above horse-head level so the animal is required to reach up to eat. Height is important so the horse does not accidentally get caught in the rack. Equine authorities can be found who say it is not good for a horse to eat from above its head;

yet horses at pasture will eat from trees that are over their heads.

Like hay racks, soft and pliable hay nets also must be hung high enough from the ground to avoid snaring a horse's foot, an occurrence of amazing frequency. A third method, feeding from the floor, is critized by some horse authorities as the "lazy" way to feed. But proponents of "floor feeding" argue that horses are digestively designed to eat off the ground.

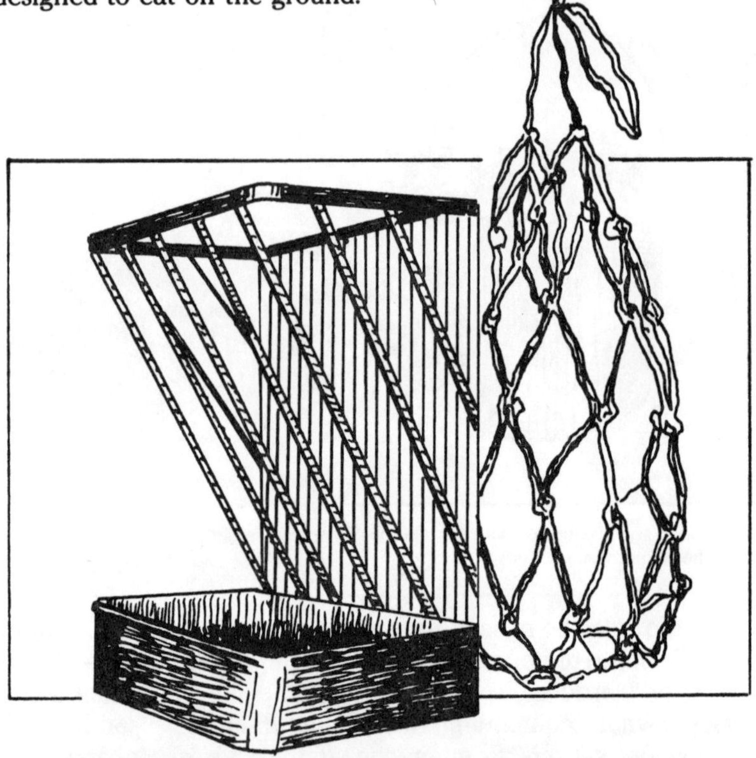

Available hay feeders include (left) metal rack with attached grain feeder and (right) hay net. All feeders should be placed high enough on the stall wall to avoid snaring a playful horse's foot.

Economy favors the rack or net method. Horses tend to strew hay about the floor. Both methods also are more hygienic than eating off the floor, considering a horse can soil the hay and reinfest itself with parasites. On the other hand, because both types of feeders must be hung high, they tend to drop hay scraps and dust in horses' eyes. Instances are known of horses being permanently

Dusty hay should be dampened with water before it is fed to the horse. Accumulation of hay dust can eventually cause emphysema, a respiratory ailment.

---

blinded while eating from these elevated containers when they accidentally jammed hay straws in their eyes.

In regards to safety, any object hung inside the horse's stall should be clipped there by a strong, double-ended snap. Snaps that close on one end but have an open hook on the other end can rip horsehide.

A horse that is standing in a well equipped, well lighted, well ventilated and dry stall, free of nails or other hazards, can be expected to feel that cozy nook is its safe haven, its asylum from the dangers of the world. Sadly, it isn't true. That equine residence may still harbor for horses the greatest danger of all: fire. In Florida and other states each year hundreds of horses die in barn fires, leading to the oft-asked question of why are horse barns so prone to fiery destruction? The obvious answer is that no more horse barns burn than other types of structures. It appears otherwise, however, because barn fires at race tracks or elsewhere often involve the deaths of many horses at one time, a tragedy that makes for memorable newspaper or television headlines. Less obvious is the fact that many barns are little more than kindling that awaits a torch — decrepit buildings with ancient electrical wiring. Add to those factors the dry wood chips or shavings on stall floors, and bales of hay strewn around, and the settings make perfect incinerators torched by short circuits, cigarettes or what-have-you. In Florida, ignorant horse owners or employees may pile up manure against the side of a barn, not knowing it can spontaneously ignite from the heat it builds up while decomposing. Improperly cured hay can build up heat and flame up just as unexpectedly.

Sprinkler systems are an obvious aid in preventing such needless equine deaths. But these are out of reach for most owners' wallets. Affordable are, however, many types of smoke detectors and fire extinguishers. Both should be placed on walls at each end of the barn and in the feed room. Detectors must be checked often,

Applying to wood a spray of finely ground garlic and hot peppers mixed "to taste" with salad oil and water, will keep a horse from chewing up its stall.

---

too, because dust and dirt can render the alerts useless. Other steps barn owners can take to prevent a flaming tragedy include:

—Use noncombustible building materials or treat the present structure with fireproofing "paint" that can be purchased at building supply stores.
—Check and replace barn electrical wiring regularly.
—Never hang bare lightbulbs in horses' stalls.
—Keep the barn area clean of old lumber and rags.
—Store gas-powered equipment and gasoline anywhere but in the barn to avoid an explosion.
—Post and enforce No Smoking signs.
—Install lightning rods.

One Florida horse owner actually holds periodic "fire drills" to work out the most efficient horse evacuation system. Of course, without smoke and heat, the simulation is not as panicky as an actual fire would be. Still, it gives her the opportunity to know which horses can be led out — and to where — most efficiently. She has learned not all horses can be led with a blindfold without some practice. The best way to shoo horses out of a burning barn is to simply open the stall doors. Some horses, however, will refuse to leave their safe haven and must be led out. Animals that absolutely won't budge should be left until all other horses are given a chance to escape. Stall latches or stall doors that are prone to jamming should be repaired or replaced before a fire occurs. And it certainly becomes impossible to get a horse quickly out of its stall if the stall is padlocked, a dangerous practice employed by some. Such imprisonments sometimes do occur. In one case, the divorcing owners were involved in a horse custody battle; in the other, the barn management locked up a horse because the owner had not paid her boarding bill. Such incidents are disgustingly dangerous for the animals should a barn fire break out.

Not nearly as dangerous but totally exasperating can be trying to trailer a horse that refuses to load. The day comes to every horse

"Anybody with an automatic waterer ought to yank it out and put buckets back in. Good horse owners know how much water their horses are drinking, and that's impossible with an automatic waterer." Dr. Ken Vasco, United States Equestrian Team veterinary advisor.

---

owner when the horse must be transported — from one barn to another, to shows, to veterinarians or to riding clinics. The key to trailering is practice. A horse that is never trailered is going to be frightened at first. All important, too, is to make the loading and the trailer ride as pleasant as possible. A horse that is beat with a whip into a trailer cannot be expected to like it. Some of the oldtime trainers, however, swear by disciplining a horse into trailers. In rural North Florida it is not unusual to see cow horses, fully saddled and bridled, hopping right into trailers. The theory, as stated by one of those cowboys, is his horses are far less afraid of the trailer than they are what will happen if they refuse the ride. But there is a better method. It is to take time during training to acquaint the horse with the trailer. Pull the trailer into the pasture, secure the doors open and make sure the trailer is well supported. Most horses will grow curious about the vehicle and sniff it inside and out. Feed the horse some meals inside the trailer. For the first few times, the horse probably will just stretch its head and neck inside; gradually, the front feet will step inside and finally the whole horse will make the move. Be patient. At each accomplishment, reward the horse with a bite of carrot or apple. The next goal is to trailer the horse safely. Regardless of the type of trailer used, it should be safe. Flooring must be checked regularly because it rots out — and a horse can fall through it. Inspect hitches and tires. Be sure brake lights and turn signal lights are operating. Metal joints should be checked for that Florida nemesis, rust.

To more safely trailer a single horse riding in a two-horse trailer, the horse should be loaded in the left-hand slot, which is the side nearest the crown on a road. If possible, preferences of the horse should be considered, too. One Florida horse trainer had a show horse that would only ride backwards in the van. Fortunately, the roomy van afforded this luxury, whereas a one- or two-horse trailer would not. Likewise, some horses do not like to be tightly confined in a trailer. Snubbing up the lead rope tightly to the trailer immobilizes the horse's head. While that's desirable in some cases, especially if a horse tends to turn around while riding, the tight headhold can panic other horses who need freedom to balance themselves. Proper trailering boils down to the same consideration a motorist would afford a human passenger. And that means taking corners slowly so as not to throw anyone off balance, a sane and safe speed, and providing a car — or in the case of a horse — a trailer, that will get to its destination without incident. True, some horses, like children, are climbers or scramblers and don't enjoy being trailered. But if an owner makes the experience as pleasant and uneventful as possible, most do not mind.

BOARDING

Many Florida horse owners, particularly those who live in the city, must board their horses in a private barns. And no matter where an owner lives in the state, or in what caliber barn the horse is kept, boarding is always a headache.

First there's the problem of choosing a barn, with finances usually playing a big part in the choice of Dobbin's diggings. The state has many boarding facilities that range from dog holes to equine Taj Mahals. And, ironically, it's not always the shabby-looking barns that are the worst. A boarder soon learns that what matters most is the barn's management.

For example, an exquisite barn facility can be owned and operated by a non-horse person. That seems often to be the case. And non-horse management is too frequently motivated by profits, not by a love of horses. As a result, stalls are not cleaned because shavings are too expensive. Horses are not fed regularly because feed is expensive. Feed is changed frequently and suddenly because a newly discovered brand is less expensive than the old brand. Hay is of poor quality and meagerly fed because of cost. Horses are not turned out into available pasture because the animals will destroy the grass. Riders are asked not to use the riding ring because the horses tear up the grass, or because riding lessons are being given in the ring — lessons that earn a kickback to the barn owner. Many urban horse owners eventually loosely interpret what that Latin phrase, "Caveat emptor" means — boarder beware. A discontented boarder can do little but move to another barn. But each new barn has its own problems. So it is best for the boarder to make a list of priorities and decide to put up with what he or she considers minor inconveniences.

That barn priority list will differ from person to person, but it should look something like this: sturdy and safe stalls and pastures, good feed and hay, good watering, accessibility to the horse, privacy from bothersome strangers, good-size riding area, clean environment, toilets, knowledgeable management, cooperative and honest management, trails nearby, good grooming facilities, honest fellow boarders, ample tack storage area, lesson availability, appearance of facility, luxuries such as snack and cold drink machines.

Naturally, sturdy and safe stalls and pastures are first consideration. The smart horse owner thinks of the animal first. Stalls should be roomy, dry and airy, with a clean water bucket and grain feeder. Feed and hay should be of good quality and fed at the same times each day. Hard as it is to believe, some barn owners do not realize the importance of feeding horses. One Florida barn owner, who never had been around horses before purchasing his large horse facility, puzzled over feeding horses every day: Wild horses do not eat often, he mused, so why do boarded horses have to eat every day? Fortunately, letting a horse totally fend for itself is not an attitude pervasive among barn owners, but it does show up in practices other than feeding. Take accessibility to the horse as another example. Sometimes horse owners must enter the boarding barn at odd hours, yet some barns lock their gates. Not even a veterinarian can get inside on an emergency call. Conversely, accessibility by the barn owner's family members can be troublesome. In one Florida barn, the owner's youngest child fed pencils to all the horses, while other family members would occasionally pluck any horse they wanted from a stall and ride it.

Also of priority is a good sized riding ring. Size is important because a ring can get too crowded with riders for everyone to pursue his or her individual interests. Worse, barn owners with too small a riding ring may expect boarders to ride in turn-out pastures, among loose and grazing horses.

The best barns have separate areas for lessons, for dressage, for longeing, for schooling over fences and for pasturing. All such areas should be free of such hazards as potholes, barbed wire and debris such as car or tractor parts.

It is also a good idea to inspect any prospective Florida boarding facility during the rainy season. So much of the state is low-lying that a light but steady two-day rain can turn a barn into an unusable quagmire for days afterward. Flooded stalls become veritable cesspools. And no horse owner should keep a horse in a leaking stall, for if a stall roof leaks, then the feed room roof probably also leaks, so foodstuff may be moldy.

Whether a facility has toilets sounds like a minor consideration, but most horse owners spend hours at the barn, so plumbing gains importance as the day wears on. Some barns offer only portable johns or old-fashioned, outdoor privies. Related to that subject is whether a facility has room for changing clothes. Horse owners often must come from school or work to ride, and it is handy not to have to stop by home first.

Of course, nothing in a barn can function if the management is not knowledgeable and, preferably, cooperative as well as honest. Knowledgeable management checks each horse many times daily and recognizes the onset of illness. It is management's responsibility to call the owner immediately if something is wrong with the horse. A good barn manager takes the initiative to call a veterinarian if the owner cannot be reached. Moreover, wise managers listen to boarders' complaints and correct problems, without playing favorites. In defense of barn owners, it should be noted that some horse owners are chronic complainers who can sour the atmosphere of even the most efficiently run barn. Another enjoyment is trail riding. But around many of Florida's cities, trails are growing scarce as communities expand, so owners who enjoy this type of riding must be prepared either to trailer their horses to trails or select a boarding barn that is near open riding areas.

Closely related to riding is grooming a horse. Most grooming chores can be done with the horse cross-tied in its own stall or in an aisleway. But there are times, especially if an owner is going to a show, that the horse must be bathed, clipped or braided. Such grooming racks provide necessary space for doctoring a horse's wounds or hosing down the horse after a ride.

Horse folk say that honesty among fellow boarders is a highly desirable quality in a barn. Indeed, no one wants to worry about a saddle or bridle being stolen. Nor does a boarder want to put up with the likes of a sticky-fingered "feed snitcher" at one barn who would enter stalls and remove any hay or grain the horses had not yet finished. Horse owners can foil such dishonest colleagues, usually by notifying management of the problem in a constructive, non-gossipy fashion.

All boarders should have a lockable tack trunk that is roomy enough to hold a saddle, bridle and grooming equipment. An option is to take tack home after each barn visit, but if that is too inconvenient then make sure a barn has ample trunk storage space that is clean and dry. Storing feed and hay with tack invites dust, rodents and insects to mess up leathergoods.

One seldom-considered aspect of a boarding barn is availability of lessons. It is always a plus if a reputable instructor is affiliated with a barn. Some barns work "deals" that give boarders a price cut on group lessons or clinics. On the other hand, some boarders prefer their own coach and will pay good money to have that person travel to the barn to give private lessons. Boarders often are shocked to learn that such "outside" instructors are not welcome — that only one person is permitted to give lessons at a barn. Not surprisingly, that "official" teacher often is the owner or manager. Or, in other cases, the barn management demands a percentage of the outside lesson price for use of the ring or the jump course. This is unfair, however, because boarders already have paid for using the facility.

Related to this situation is whether all farriers and all veterinarians may come to a barn. It always suprises boarders to learn that their favorite farrier or doctor cannot practice there because the barn has its own personnel — or uses just one farrier or veterinarian exclusively. Boarders encounter this dilemma especially if they move into a barn owned by a farrier or a veterinarian.

Far down on the priority list should be the appearance of the facility. Of course, the barn should be clean and not a public embarassment. But manicured landscaping, showy barns and lounges do not necessarily mean the barn management is competent. In fact, an overly ritzy facility can indicate the emphasis is on glamour and not on quality feed, hay and bedding.

Finally, check out the availability of such amenities as food and drink vending machines. Not all horse people want to pack a picnic cooler on the days they spend at the barn, or drive to a restaurant for nourishment. The negative of such machines on the premises is trash and leftovers being strewn about the barn, attracting insects and rodents.

Most barns that Florida horse owners inspect offer only one type of boarding plan, full board. This means boarders pay a monthly rental fee and receive a stall, horse feed, hay, water and use of all facilities. Full board means, too, that the horse's stall is cleaned at least once a day and that the horse will be fetched and held by barn personnel when the farrier or the veterinarian call. Full board is a godsend for parents of child riders or adult owners who work. But it is the most costly boarding arrangment, running from $100 a month and up.

Alternative boarding plans such as semi-partial or partial board are available at other barns. Semi-partial board, a blend of full board and partial board, usually gives the horse owner use of the riding and grooming facilities, a stall, stall cleaning service and, in some cases, barn personnel will feed the horse whatever grain or hay the owner leaves in a pre-designated place. One advantage to this system is that the horse owner, by purchasing feed and hay, chooses the quality and quantity to be fed but does not have to do the feeding. There also is the advantage of lower cost over the full board plan. The horse owner pays a reduced rate because he or she is purchasing feed. In Florida, semi-partial board rates are about $85 and up.

The bare-bones boarding plan, partial board, means the horse owner is receiving only a stall and permission to use the riding and grooming facilities. Owners are required to purchase their own feed and hay, must feed their horses and must clean their own stalls. This plan generally costs $50 or $60 a month. It saves money but more importantly, the partial board system permits the horse owner to regulate the horse's feed and stall cleanliness. Such horse care is time consuming, though, and working adult owners, or parents whose children ride, find themselves out at the barn seven days a week, often until late at night.

Boarders can circumvent this matter of long, daily barn chores by joining with other semi-partial or partial boarders to take turns feeding and cleaning. It should be noted, however, that this type of boarding requires that horse owners buy some feed and equipment storage unit — usually a portable shed. Barn owners usually permit such sheds near stables, although the many different styles, sizes and quality of the sheds can soon resemble a shantytown. Moreover, barns that offer such flexible boarding plans usually have designated stalls or barns for the different pay scales. As such, partial boarders may end up with the shabbiest stalls. It is growing more difficult to find Florida barns that offer a variety of boarding plans and prices. Barn owners are learning it is less trouble to offer only full board. That way they know all horses are being fed and all stalls are being cleaned. Moreover, full board arrangements cut down on boarder squabbles that occur when high-paying full boarders believe low-paying partial boarders are hogging the facilties.

Similarly, it is increasingly difficult for horse owners to find a barn in Florida that offers a written, rather than a verbal, boarding contract. Those barns that do put everything in writing offer contracts often brimming with musts for the boarders and few responsibilities for the barn owner. Thus, if boarders are abused they have little recourse but to move out.

Without a written boarding contract, a priority list made out by the boarder as mentioned above takes on added significance. Prospective full boarders particularly should ask the nitty-gritty questions, and make a thorough inspection of a barn and its facilities, before agreeing to move in. Those questions should include: What is fed? When? What is fed if regular supplies are not available? How often are stalls cleaned? How often are stalls dug out? Is bedding available if previous supplies are exhausted? Are all horses turned out daily to pasture? Does a boarder have access to the premises at all hours? What about bringing in a teacher for lessons? Can a boarder choose his or her own farrier and veterinarian? Is the barn owner or some other responsible person on the property at all times? Is the barn a fire trap? Don't just ask what is the boarding fee. Ask whether boarders are expected to provide their own feed or bedding if the barn's supplies run out. Check, too, if there is a reimbursement policy in case boarders must supply these items for which board is being paid.

When visiting a prospective facility, go several times. Go at night and ensure the barn and ring are well lighted. If pasture is offered as one of the amenities, check it to make sure it is not just a sand pit. Does the facility have a reputation for having escaped horses? Look at horses that have been at the barn for the longest time — are they healthy? Have horses died at that barn? These are tricky inquiries. Barn owners are not necessarily at fault because some horses are in poor physical condition or die. Still, an unusual number of equine deaths or near-misses should raise a red flag. In

Florida heat requires extra salt intake for horses. Always provide salt blocks. Regular (white) and mineral (red) blocks should be placed in stalls and pastures. Be sure to shelter blocks in the pasture from rain.

---

this regard, try to verify rumors — a barn owner may have been unjustly blamed for a horse's death or an injury.

Surprisingly, present boarders are poor information sources about a facility. Current boarders may be friends of the owner or may fear being kicked out of the facility if he or she gives a bad recommendation.

Good sources about barns are blacksmiths and veterinarians. They visit many barns every day and may recommend one from at least a hygienic standpoint. Query local horse clubs, too; they occasionally keep track of good or bad boarding facilities in their area.

FEEDING YOUR FLORIDA HORSE

It is a proven fact that there is no substitute for properly feeding a horse. A balanced diet of protein, vitamins and minerals does its part to keep the horse not just healthy but performing or breeding to its top potential.

A first-time Florida horse owner soon will find that horses, like most humans, acquire a "live-to-eat" philosophy. In fact, food is probably the most powerful influence we have upon our horses' lives. Thus, feeding can become a valuable training tool. Many an owner has summoned a pastured or loose horse with the gentlest rattling of the grain bucket. Likewise, a handful of grain can reward good behavior or induce a stubborn equine to clamber into a horse trailer.

Feeding a horse, however, is unlike feeding any other domestic animal. Horses have, it seems to most people, very odd digestive systems. For example, despite its size, the horse has a small stomach with only about a two- to four-gallon capacity. Furthermore, horses' digestive systems are easily turned topsy-turvy. Contaminated foodstuff that would prompt most animals, including humans, to get sick to their stomachs, causes other problems in horses because equines do not vomit. Instead, they may colic or suffer with laminitis. Colic, a digestive upset, can arise from changing grains quickly, from feeding contaminated grain or hay, from riding too soon after feeding, from feeding or watering an over-heated horse, or from the horse "bolting" or eating too fast. Similarly, such foot problems as laminitis (founder) can occur from over-feeding or from feeding tainted grain.

Despite the fact that equine nutritionists call feeding a science, opinions seem to differ among veteran horse owners as to what is the best fodder. Perhaps only horses know. Each animal is an individual, so what is gourmet heaven for one horse may gag a stablemate.

Most equine nutritionists agree on one basic fact: Be consistent. When you find a good quality ration that a horse eats happily, stick with it. Horses have a limited palate. Many novice owners mistakenly launch a "salad bar" feeding system. The owner rushes out to purchase a bag of this, a bag of that. Not only is this costly, but it may produce an equine epicurean who is finicky enough to shame Morris the Cat. Galloping gourmets are made, not born.

Likewise, the horse should be fed at the same times each day;

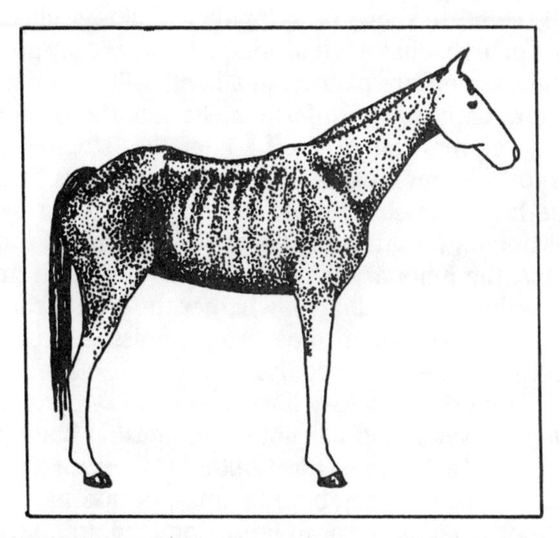

Severely malnutritioned horse. Younger horses, under five years, are more likely to recover, although they may never achieve their full potential. Older horses, in their teens or older, may never fully recover their prime weight.

fussy eaters or high-strung nervous animals can grow overly agitated while waiting for a late-arriving meal. Also, avoid stressing a horse at mealtime whenever possible to prevent colic. There is another reason to settle on one good feeding regimen — an uneaten meal is often the first clue an owner has that ol' Dobbin is feeling puny. An animal that has gobbled its feed for months, then goes suddenly sour on its diet, may be communicating either that it is sick or that something is wrong with the feed. The owner should inspect it. Does it look like it should? Is it buggy? Does it smell fresh and free from mold? Many horse owners believe a horse knows if its feed is bad and so will always refuse it. That's not true. Young horses, or especially hungry ones, will dive in and scoff their fill before so much as sniffing it. Owners should always check the aroma and appearance of any feed or hay before giving it to a horse.

True, most horses will eat their food even if it contains a few stray ants. On the other hand, if feed is being carried off an oat at a time by a colony of ants, especially big red ones, the horse likely will walk away from the bucket. And, understandably, most horses flatly refuse to share their meals with cockroaches or rats.

Occasionally, a horse will snub its feeder for no apparent reason. If the horse shows no symptoms of illness, check its water bucket; a horse that has not had adequate water may go off its feed. Automatic waterers are handy, but a horse will not drink from one that is slimy with algae. Unfortunately, too, the owner who has automatic waterers cannot tell if a horse has stopped drinking.

The most common cause of not eating from its bucket occurs when the horse is being turned out on new spring pastures. No longer as hungry for grain, the newly pastured horse should be fed less rations, the amount dependent on how much it finishes in a reasonable time. Depending on whether the feed is commercially-prepared or whole grains, most horses finish eating within two hours. Some trainers expect the plate to be cleaned within 30 minutes. If a suddenly finicky horse shows no symptoms of illness, is drinking its water, and has not spent unusual hours at pasture, then check with the feed distributor. Makers of commercially prepared rations, even top brand names, occasionally change an ingredient or make an error in preparing feed. In one Tampa Bay area instance, the "binder" ingredient that keeps pellets from being powder was accidentally left out by the mill. Though the omission

"I think commercial feeds are the best way to go. People get to mixing and matching their own custom formulas and they run out of something, so they substitute something else. The next thing you know, they've got their formula all messed up and they're having trouble with their horses."
Dr. Edgar Ott, Equine Specialist, University of Florida

---

was no threat to life, it changed the feed's taste. Horses everywhere were snorting indignantly at their feed buckets. The local distributor gladly exchanged old foodstuff for bags from a new shipment of tastier fare. Equine nutritionists and veterinarians consistently say the average horse owner tends to overfeed his or her steeds, that many Florida horses waddle around with unnecessary heat-producing fat. It's common, say veterinarians doing a post-mortem examination on a dead horse, to find huge blobs of fat around the animal's heart. Competition horses in Florida should favor the lean side of the scale. Halter horses usually are shown carrying a few more excess pounds. But many halter exhibitors believe heftier is better — to hide conformation faults. Good judges can see through that ruse and are increasingly ranking fatsos below the more fit competitors.

Nutritional Requirements

Feeding a horse can be as simple or as complex as the owner wishes. Some people opt to add all sorts of goodies — vitamin supplements, wheat bran, linseed meal, ground limestone, molasses. Some of these additions to basic grains such as oats and corn will boost protein content. Other additives, such as bran, are thought to have a laxative effect.

General thinking among equine nutritionists is that mature horses, or those in competition or in early pregnancy, do well on 10 percent protein. Stallions, horses in training or in late pregnancy should receive 12 percent protein. Yearlings and nursing mares should receive 14 percent protein. Nursing foals and weanlings six to nine months old should receive 18 percent protein.

A variety of grains are available in Florida to be mixed together, either by the owner or by select feed dealers, into a "customized" feed ration. Or, many horse owners opt to mix these grains, such as oats or corn, with already mixed commercial feeds.

Oats — rolled, crimped, crushed, cracked or whole — provide roughage combined with a slow release of energy. Whole oats are preferable if they are fresh (shiny, pale yellow) and free of dust or contamination. A wide variation in oat quality exists because of the difference in hull (the "shell" of the oat) content. Oats generally contain about 10 percent protein. Corn is high-energy food and is popular in cooler climes. In Florida, horse owners shy away from corn because it is a "hot" feed, that is, it has high energy content per unit of weight. But a Florida horse that is worked daily should not have any problems with some corn in the feed. Conversely, idle or overweight animals should be fed corn sparingly. Straight corn rations are never recommended. Processed corn (cracked or finely ground) can pack down in a horse's digestive system and beckon colic.

Recommended Custom Feed Ration For An Average Pleasure Horse

Ingredient	Percentage
Oats	40.0%
Corn	40.0%
Molasses	8.0%
Wheat bran	5.0%
Dehydrated alfalfa	5.0%
Salt	0.75%
Ground limestone	0.5%
Premixed vitamins and minerals	0.5%
Dicalcium phosphate	0.25%

Advantage of feeding a custom-mixed ration is that it allows a horse owner to quality inspect each ingredient. An owner knows rather precisely what the horse is eating and can make subtle protein or vitamin adjustments. The disadvantage is that many ingredients must be purchased and mixed. Moreover, figuring ration percentages and mixing can be time consuming in an owner-operated multi-horse barn. Also, storage is more space consuming.

Much easier is to purchase a good quality, commercially prepared feed. There are many brands on the market and nearly all offer a range of 10, 12 or 14 percent protein. The feed comes in "sweet" varieties in which corn, oats and pellets are coated with molasses. While horses generally like sweet feeds, there is quite a difference in consistency from brand to brand. Some are almost gooey-moist; others are dry.

Prepared commercial feeds also are sold in pellet form. Again, horses generally like these pellets and they are handy to use. Pellets come in different mixes, ranging from "complete" to "grain" to "supplement" rations. The complete pellet is just that, and includes hay, grain and supplements; grain pellets are pelletized grain; supplements are pelletized protein, vitamins and minerals.

Commercial pellets and sweet feeds have pros and cons. As mentioned, the primary plus is their convenience. Sweet feed does attract a few more insects than pellets do. On the other hand, if the horse dribbles water in its feed bucket, pellets can melt and become a soupy swill. With both sweet feeds and pellets, a horse owner must trust the milling company's protein and ingredient label more than with custom-mixed rations. Unfortunately, horses often plow through their meals of commercially prepared feed much faster than they eat whole grain feeds, which must be chewed more thoroughly. The fast eaters, with more leisure time, then sometimes turn to chewing wood or indulging in some other unsavory vice.

Many horse books offer charts and graphs on amounts to feed horses based on "per 100 pounds of bodyweight." The charts are ignored, understandably, by most backyard horseowners who have neither the patience nor the weight scales to precisely measure their horses or the percentages of feed. At the same time, baffled neophyte owners should never hesitate to ask a veterinarian, a breeder or a trainer about the feed needs of an individual horse. Horse folks love to give advice.

Inasmuch as some prepared horse pellets contain hay, and cubes made of hay also are now available, some owners no longer buy and store traditional hay bales. Owners who do favor feeding baled hay should feed a portion of each day's forage with each meal. Hay portions are called "flakes." A larger flake is usually fed with the evening meal. It can be amusing to observe veteran horse owners new to Florida trying to figure out the Sunshine State's hay offerings. Accustomed to such "northern" hays as timothy and alfalfa, the newcomers are totally unfamiliar with Florida's Coastal

Bermuda grass, Suwannee Bermuda grass and Pangola hays.

Coastal hay probably is the state's most popular. It was developed in Georgia and is common in Florida because it is hardy in hot temperatures, in occasional winter frosts and in the state's drought-to-hurricane rainfall extremes. Alyce clover is the state's only legume sold as feed. Legumes are not grasses but are plants such as beans or peas, which can be fed as a supplement or as a substitute to grass hays.

Florida farmers have been experimenting with a new Florida alfalfa hay, but reviews have been mixed and it is not yet plentiful enough to be economically competitive with Bermuda grass. Horse owners who yearn for timothy or alfalfa hays usually are delighted to find such hays are trucked into Florida regularly and available much of the year. Unless the owner has a large farm and can afford to buy a truckload, however, the retail prices of these imports at most local feed stores is dismaying — often two to three times higher than prices paid per bale up north.

Any hay, whether native or imported from up north, must be fed when it is fresh. It should be greenish-colored and smell sweet. It should be free of heat, of insects such as blister beetles, of foreign objects such as pieces of metals or glass. Some weeds are bound to get into the baler, but thorny brambles should be removed. In fact, it is good practice to remove anything that is not hay, just in case it is poisonous or could upset the horse's stomach. Also, hay is like any harvested food — it begins to lose its nutritive value as it gets older. For this reason, some owners prefer to buy hay by the dozens of bales rather than by the ton. This practice is less economical, but keeps fresh rations arriving more often.

Horse owners should use care with Coastal hay. Occasionally it is very thinly cut. Some veterinarians believe such fine hay strands may pack down in the horse's stomach and cause colic.

Storage of hay should always be in an airy but perfectly dry area. Criss-crossing the bales on top of each other provides more air circulation than straight stacking. The bottom layer of hay bales should never be placed directly on the ground or they may be spoiled by dampness before they are fed to horses. It is debatable as to whether horses really need hay, but it is certain that they enjoy eating it. Stalled horses especially seem to crave the chewing that hay provides.

Horses also enjoy eating treats. But a treat is defined in equine terms as a carrot or an apple. Several carrots and no more than two apples are standard daily fare. Many novice owners think it is cute to ply Dobbin with soda pop or a candy bar. Or, they reward their horses with that fabled treat, a sugar cube. But sugar causes tooth decay, and an irritable 1,200-pound horse with a toothache is not cute. Boxed treats, similar to those for dogs or cats, are available for horses. They come in different flavors such as apple or carrot, and are handy to feed. Such treats are rather costly, though, and are subject to mold and mildew if not stored in a jar or plastic container. Owners should also be careful about how much protein a horse is fed. As mentioned, the standard percentages range from ten to eighteen, depending on the animal. Ultra-high protein supplements (25 percent or more) can make a high-strung horse absolutely unmanageable. In fact, some equine delinquents become four-footed saints simply by lowering protein levels. Make any changes gradually.

Other feeding tips that apply to Florida horses include:

— Feed frequently. Three small feedings usually wastes less feed and the horse digests more nourishment than from one or two larger feedings. Some racehorse training barns advocate six to eight small feedings each day.

— Buy good quality feed. Money will be saved in the long run in better nutrition and the resultant fewer medical problems.

— Store feed in rodent and waterproof containers such as galvanized garbage cans. Leaving feed in the bag invites mold, insects and rats.

— If possible, separate horses at feeding time to keep the dominant animal from hogging the trough.

— Inspect the horse's feed bucket. Be sure it is clean, free of insects. After the horse has eaten, check the bucket for dropped grain. If a horse seems to waste a lot of feed, have the animal's teeth checked for a chewing problem.

— If a horse is laid up by an injury or its work is reduced, cut back on feed gradually and slightly increase the hay ration.

— If a horse has been given a good chance to finish its meal but does not, remove it. Old feed contaminates quickly and attracts roaches and rats.

Learn what mildew or mold looks and smells like in grain and hay. Owners who are ignorant of these two Florida nemeses may as well feed their horses poison. Contaminated feed causes colic,

Get rid of disease-breeding manure piles by giving it away (or selling it, if the market will bear) to local gardeners or farmers. Hang a sign out in front of your barn or run an inexpensive ad in the local newspaper.

---

the leading killer of horses. Naturally, what goes in one end eventually comes out the other. Check the horse's droppings. A healthy horse's "meadow muffins" that contain undigested feed indicates the animal is not chewing its food thoroughly. While this can signal a tooth or mouth problem, it may also indicate a greedy horse who gulps down the food before it has been chewed. Greedy horses invariably colic more often than slow eaters. Droppings should be neither rock hard nor "cow patty" soft.

In addition to colic or laminitis (founder), other diseases such as rickets or "bran disease" can result from improper feeding. But with improvement in equine nutrition research, better quality feed and supplements are making such ailments rare.

FLORIDA PASTURES

In Florida, keeping spaces grassy green for horses is every owner's chief challenge. No matter how hard the horse owner works, it takes only two weeks without rain to make pastures change magically to sand. One of the keys to battling Florida's sugary sand is not to permit pasture to be overgrazed. Keeping too many horses on too little land is a common failure arising from the misconception that two or three horses per acre will not eat the pasture bare. Florida agronomists recommend keeping a maximum of one horse per acre. In many areas, keeping no more than one horse per two acres will help to ensure that pastures stay edible during periods of drought, which in Florida, generally occur in the spring and fall. Such recommendations are conditional, of course. For example, more horses can be turned out on each acre if they spend only an hour or two out-of-doors. And tolerance of pastureland to grazing also depends on the quality of the soil and the grass.

With enough liming and fertilizing, even the poorest Florida sand can support a minimum number of horses. Throughout most of the state, soils are some type of sand — loamy, well-draining,

Florida's gopher tortoise, whose burrows dot horse pastures, is a species of special concern and may not be killed or captured except in a few areas of the state. It is illegal to pour toxic or flammable substances down gopher burrows because other important Florida animals share burrows with tortoises.

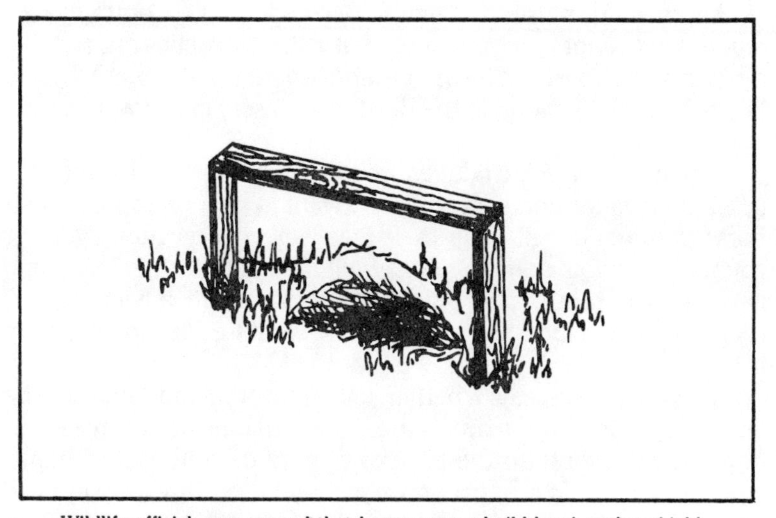

Wildlife officials recommend that horse owners build barriers that shield horses from stepping in gopher holes but at the same time permit burrow tenants to come and go at will. Simple and inexpensive barrier that has been tested as effective is illustrated.

excessively-draining or poorly-draining. The best of these sands grows some hardwood trees and, in general, a minimum of scrub. The poorest dry sandy soils support only the barest native vegetation such as wire grass and palmetto scrub. Equally poor but swampy sandy soils foster Florida's shallow ponds and mosquitoes. Several exceptions to this sandy environment do exist. One is in south Florida, near Lake Okeechobee, an area featuring high contents of peat and muckland. Much of this land has been drained but is quite rich and supports pasture very well. Similarly, north Florida's Panhandle and border areas near Alabama and Georgia have clay soils that tend to hold water and dry to a hard consistency.

Grasses that grow on most of Florida's pastureland have been imported from somewhere else, usually South America or Africa. Some of it, such as digit grass, is not cold tolerant and therefore suited only to south Florida. Digit grasses include pangola, slenderstem and transvala.

A popular pasture grass imported from South America is Bahia grass — either Argentine, Pensacola, Paraguay 22 or the Common variety. Bahia grasses prosper all over Florida and withstand heat, drought and cold. These grasses have been cultivated as hays.

Another grass used for hay is Bermuda grass. Common Bermuda grass is not recommended, but hybrid varieties are suitable for pastures. Coastal, Suwannee and Coastcross-1 all are hybrid Bermuda grasses. None of the Bermuda grasses grows well in very poor soil.

New grasses are introduced frequently in Florida. Therefore it is a good idea for horse owners who want to start or improve their horse pastures to call upon their local soil conservation agent or county extension agent. Besides being familiar with existing and new grasses, these state officials can give good advice on soil type and will analyze soil to determine what fertilization program is best.

Analysis also reveals whether soil (or pasturelands) needs to be limed. Most do. But the amount and type of lime varies from area to area, depending on the type of pasture desired. For example, clover demands an acid pH between 6 and 7; pasture grasses need only 5.5 to 6.5. Soil experts stress that pasture nutrients such as calcium and magnesium also need to be corrected with additives.

Most commonly added to pastures is fertilizer. Just what kind

A horse medicine cabinet should be part of every barn and should include: Baking soda - Nitrofurazine spray or ointment - Iodine - Aspirin - Milk of magnesia - Alcohol - Mineral oil - Phenylbutazone tablets - Cotton gauze squares - Stretch "Ace" bandages - Surgical soap - Liniment - Adhesive or electrical tape - Thermometer - Scissors - Powdered wound dressing - Sponges - Measuring cup - Petroleum jelly.

---

and how much again depends on the type of pasture grass desired. New plantings should be treated differently from established pastures. For instance, new pastures might be fertilized with 10-10-10, that is, 10 percent nitrogen, 10 percent phosphorus and 10 percent potassium (potash). For an already established pasture, however, one that has not been fertilized for several years, agronomists recommend a 15-5-10 fertilizer. Again, individualized recommendations can be obtained, free, from county extension agents.

KEEPING YOUR FLORIDA HORSE HEALTHY

Nothing quite catches the eye like a shiny reflective coat of hair on an animal. In horses, as in all animals, such an attractive coat is due only partly to proper grooming. The horse must be healthy inside, too, or no amount of brushing will promote a satiny outside appearance. One key to such equine health is avoiding parasites. In Florida, where the air is warm and humid nearly the year round, that can be a constant challenge. The state has an unusually long "growing season" for critters that fly and crawl their way onto and into other living things. With horses, most of the fliers remain outside the horse, on its coat. There, they bite and cause the horse not only discomfort but also allergic reactions. Worse, the fliers lay eggs that turn into crawlers that end up inside the horse. These larvae then feed on the host animal.

Internal Parasites

Parasites of any kind are no respecters of equine beauty, value or strength. They are totally promiscuous. And it is the pests that infest the horse's internal organs that most concern owners and veterinarians. Such parasites rob the horse of nourishment. They also decrease the animal's work efficiency and cause digestive and breathing difficulties. Complicating matters is the sheer number of such parasites. Equine medical experts estimate horses are affected by fifty to eighty different internal parasites — more than affects any other domestic animal. Staggering as those numbers are, most Florida horse owners need be continuously alert to only a portion of internal pests. These few not only are dangerous but can be fatal if left to become a heavy infestation. Chief among them is a variety of strongyles, which come in two basic sizes, large and small.

Large strongyles, most veterinarians point out, can be devastatingly harmful to horses. Adult strongyle worms live inside a horse's intestinal tract where they lay eggs that pass through the horse's body to the outside. If weather conditions are favorably warm and moist, usually the case in Florida, the eggs hatch and develop into larvae. Larvae inevitably are eaten by another horse, or reingested by the original host horse. Larvae then go to the intestinal tract and enter its blood vessels. These circulatory "turnpikes" carry the larvae to the horse's liver, heart, spleen, lungs and other organs, where damage is done in a variety of ways. Larvae inflame blood vessels, which alters blood flow; weakened expansions called aneurysms may form that can later burst; abscessed, that is, infected areas, may form that weaken the horse's system. Stricken horses generally are anemic and emaciated, exhibiting dull coats and listless behavior.

Cousins to the large strongyle are about 30 different species of small strongyle worms. The small worms, however, are not as dangerous. For instance, their larvae tend to remain in and damage only the intestines.

Yet another wormy pest — one of Florida's most apparent — is horse bots. The bots are actually larval stages of bot flies, which attach themselves on a horse's stomach lining. Found throughout the United States, bots are particularly bothersome in Florida because they are nurtured by the warm and humid climate. For example, bot flies lay eggs for just four to six months in northern climes but enjoy ten months (March through December) of activity in Florida.

The female bot fly is black and yellow striped, about the size of a honey bee. One female fly can lay 500 eggs on a horse's body hair, usually around the front legs. Several species of bot flies exist in Florida, including throat and nose bot flies, which lay their eggs in those areas of the horse's body, including the horse's mouth. First clue that a horse has bot contamination often is the appearance of small eggs on the horse's coat. These eggs can be so numerous that the horse looks sprayed with ivory-colored flecks smaller than the printed letter "o." The sticky eggs hatch for up to three months, plenty of time for a horse to get around to licking eggs off its own skin or off the coat of a buddy.

Once inside the horse's mouth, these larvae hatch and burrow into the horse's tongue and gums where they stay about a month before they are swallowed. When they reach the horse's stomach, bots hook onto the wall and await adulthood (about 10 months). Mature bot larvae pass from the horse's stomach onto the pasture or stall where they become pupae, which become adult flies, and the cycle repeats. Left unchecked, bots cause digestive problems, even colic, and are suspected of causing stomach ulcers.

Whereas bots attach to any age horse, another typical Florida internal parasite, the ascarid, a large roundworm, primarily threatens suckling and weanling foals. Fortunately, susceptibility to ascarid infection is thought to diminish as horses mature. However, for infested animals, the ascarid does its damage. An infected horse tires easily, may cough, suffers digestive problems and loses weight. Ascarid infection has been linked to pneumonia bronchitis and intestinal blockages. Infection occurs when a horse accidentally eats ascarid eggs off an infested pasture or a stall floor. Just as ascarids prey on foals, so do strongyloides, also known as thread worms. Newborn horses may be infected through their mothers' milk, particularly the colustrum (first milk). As with other internal parasites, the adult strongyloides lay eggs in the intestines. The eggs pass into the pasture or stall floor and develop into infective larvae that can lay their own eggs, which in turn develop into more infective larvae. Larvae may re-enter the horse if they are eaten by the animal or if they burrow into the horse's skin. Once inside the horse, the worms pass into the intestines. Not much is known about the internal migration of strongyloides, but it is thought foals develop a form of immunity, for few young horses past six months of age are bothered by these parasites. The few adult horses who continue to be strongyloide hosts suffer few health effects but can be reservoirs for future foal contamination.

Florida's remaining major internal parasites are more of a nuisance than they are life-threatening. For instance, some horses can be seen rubbing their tails on a stall wall or a pasture fence. This itching can indicate pinworms. The pinworm lays its eggs around the hindquarters under the top area of the tail. While the pinworm is little more than an irritation, special attention should be paid to the rubbed areas. Florida heat and humidity can prompt a small scrape to enlarge to dinner plate size in a matter of days. Flies and dirt can cause infection.

Unfortunately, many of these parasitic denizens cannot be seen. And that's where good horse ownership comes in. The keen-eyed Florida owner may not see the creepy-crawly culprits, but will learn to recognize their handiwork such as a dull coat, a distended or "grass belly," lackluster eyes and general lack of stamina and zest. Contaminated manure in dirty pastures or stalls can cause the Florida horse owner untold grief. In other words, basic good hygiene is vital to holding "bugs" at bay.

Following are some tips to keeping a horse in Florida free of internal parasites:

— Rotate use of pastures. This helps to prevent reinfestation of the horse.

— Mow and harrow pastures regularly to reduce chances of reinfestation.

— Remove manure regularly for composting. Avoid just heaping it up because flies of all kinds like to breed in manure piles. (Never pile manure close to a building; it builds up heat and can cause combustion.)

— Alternate on pastures, if possible, cattle or sheep with horses. Cattle and sheep parasites do not bother horses and vice versa.

— Feed horses from hay racks and grain feeders to prevent ingestion of parasite eggs. This is vital in a stall where the horse has a limited area to deposit manure.

In the case of bot eggs, they can be removed from a horse's coat in several ways. If the coat is heavy, say, just after winter, the eggs can be clipped off. At other times, a warm water rinse (120 degrees F) may tempt the eggs to hatch prematurely and die. In contrary cases, a small square of sandpaper can be used to buff off the sticky bot eggs.

No method of control is as effective against internal parasites, however, as a good deworming program. Horse owners new to Florida often are taken aback by how frequently horses in Florida are wormed. It's a case of taking the bad with the good — the state's perennial sunshine and warmth quickly breed generations of parasites that require constant attack. To ensure that a horse enjoying the state's subtropical climate is not debilitated by strongyles, bots, ascarids or other such parasites, frequent deworming is a necessity.

A veritable chemical buffet of worm killers are available. To make matters even better, most veterinarians pride themselves on keeping up on the best, and ever-changing, deworming treatments. Be prepared, however, for no two equine practitioners, trainers, owners, breeders or references to agree on how often horses should be dewormed. Ten years ago, almost everyone opted for two or three times a year; now, some owners are administering dewormer every other month. New studies reveal some parasites are most active during Florida's balmy winter from November to March and recommend monthly treatment with a dewormer that contains ivermectin. Each owner will have to work out a schedule with his or her equine veterinarian.

But there is no disagreement over the fact dewormers must be gotten inside the horse to where the worm lives. The two usual ways to accomplish this are: tube worming or paste worming.

The surest method, tube worming, is to pump the compound through the horse's nose directly into the animal's stomach. This method, while most efficient, has some decryers who dislike it because (1) some horses may get a bloody nose from broken nasal blood vessels, and (2) it costs more than paste worming because tubing must be done by a skilled person, preferably a veterinarian. If it is not done by somebody who is skilled, the horse could be killed should the tube be accidentally threaded into the lungs instead of the stomach.

Paste (or pellet) dewormers are a less expensive, more convenient way of controlling parasites. In recent years, these do-it-yourself products have been greatly improved. But they have drawbacks, too. For example, if a horse owner does not know a horse's weight, too little dewormer may be given. Many dewormers come in 1,000-pound-horse doses, yet the earmarked horse may weigh more than 1,000 pounds.

Some horses spit out dewormers faster than Goose Gossage hawks a chew. Still other horses acquire a watchmaker's skill at separating pellet dewormer from iotas of grain. All these factors affect whether the horse receives full benefit from a dewormer.

Other measures can keep internal parasites from getting too

Tips and Quotes!

The Healthy Horse's Vital Signs
(Adult horse under normal resting conditions)
Pulse: 34-44 beats per minute
Respiration: 8-16 breaths per minute
Temperature: 99-100.5 degrees F
In addition, capillary refill time — the length of time it takes for the horse's gums to turn pink after pressure is applied to them — is about one second. Gums, and membranes of the eye and nose should be a healthy pink.

Dangerous Vital Signs
Pulse: 55-90 beats per minute
Respiration: 25-35 breaths per minute (or higher)
Temperature: 105 degrees F or higher
Capillary refill time: 2 seconds or longer
Note: a steadily falling temperature may indicate onset of shock.

---

fast a wormhold. One is to deworm all animals at the same time and arrange for newcomers to be quarantined until they are dewormed. Likewise, visiting equines should be kept off the feeding pastures.

In Florida, young horses should be first wormed at eight weeks and again every one or two months to control the large roundworms (ascarids). Mares who are nursing foals can be wormed one month after they give birth.

It is recommended a horse owner vary the product used in a deworming program because parasites may grow immune, or resistant, to some deworming chemicals. Names and advice pertaining to specific types of deworming products are available from any equine veterinarian.

Some horse owners continue to balk at a regimented deworming program, even though evidence is accumulating that worms also can cause colic. Indeed, years ago some equine experts saw little value in such a treatment, contending preventive measures were adequate. If a horse did show blatant signs of infestation, they often administered a homemade concoction of linseed oil and turpentine or some other such readily available ingredients. Inasmuch as horses dislike being forced to consume bad tasting medicine (who doesn't?), and some owners object to the tube worming procedure, many Florida horses are improperly wormed. Unfortunately, this ignorance, neglect, fear — call it whatever — results in needless and agonizing equine deaths every year.

External Pests and Parasites

If there is any drawback to owning a horse in Florida, particularly in summertime, it is the presence of insect pests.

Some of the varmints buzz and bite both rider and horse; others, however, are quiet, small and barely let on they are busily causing welts and other problems. Most obvious are a bevy of different flies: horse, house, stable, deer, and the already mentioned bot.
often surprisingly large (about one inch long) biters that resemble giant house flies. During Florida's summer, horse owners may spy several bleeding spots on a horse, probably from horse fly bites.

A smaller cousin of the horse fly, the deer fly, is only the size of a house fly, but sports a yellowish colored body and a horse fly-size bite. Both flies are daytime feeders and deposit an anti-clotting substance that keeps wounds drippy. If a horse has many bites, if the weather is particularly hot and humid and if the horse is not kept clean and in clean surroundings, a secondary infection can erupt from these fly bites. Diseases thought to be caused by horse and deer flies include Equine Infectious Anemia (also called swamp fever), anthrax and tularemia.

Unfortunately, Florida is perfectly suited for such flies because their larvae thrive in moist areas. Moreover, these pests are often so persistent that few repellents give much relief. Adding further venom to the fly-by-day subject is the stable, or dog, fly, which also looks like a house fly. Stable flies breed in old soggy hay, grass clippings or in piles of damply rotting cuttings. Stable flies can be somewhat controlled by keeping potential breeding areas clean. However, this species is believed to roam up to 80 miles from its breeding site so no measures against it are 100 percent effective. Next on the least wanted pest list would have to be Florida's pervasive mosquito. While jokes abound about the size and ferocity of the state's "official insect" (an honor others say more rightly belongs to Florida's cockroach), the mosquito has never been a laughing matter. Millions of dollars have been spent to rid the state of the malaria and yellow fever bearing mosquito populations.

The mosquito has proven a formidable enemy to Florida's horses. The insects are known to transmit such equine ailments as Eastern and Western Equine Encephalitis (EEE and WEE). Mosquitoes also carry Venezuelan Equine Encephalitis (VEE), but its incidence is rare in Florida. Many sources also blame mosquitoes for spreading Equine Infectious Anemia.

The wise Florida horse owner regularly inspects for any mosquito nurseries such as idle or rarely cleaned water buckets, troughs or feeders.

A lesser threat than the mosquito or biting flies is the ordinary house fly that is so prevalent in horse barns. "Annoying," many horse owners would say, "but relatively harmless; their worse offense is frolicking in the manure pile." Wrong. Ordinary house flies, if numerous, can worry horses sick. What's more, house flies do transmit disease such as "pink eye" and the *Habronema* stomach worm. One study revealed more than 100 disease organisms on the body of a typical house fly. Spray systems often are seen in the posh Florida barns, systems that emit a jet of insecticide on a timed basis. Another group of innocuous-seeming insects fall into the "mini insect" category of "no-see-ums," "punkies," "biting midges," "sand flies" or, more generally, "summer itch." Research reveals these several insect species are largely responsible for itchy equines. Such small insects are all over Florida, especially in the swampy areas, and are part of the insect world's "night shift," feeding from twilight to shortly after sunrise. They breed in ponds, irrigation canals, water troughs, barrels and in moist, organic soils.

Not quite as small but just as virulent are Florida's blood-sucking horse ticks, the hard tropical tick and the spinosed ear tick. Both ticks can damage a horse's inner ear. The tropical tick has been traced to Piroplasmosis (tick fever), which can be fatal about 10 percent of the time. This villain has been reported all through Florida, but officials say the insect has a low winter survival rate north of Orlando. Recommended treatment for both ticks is dusting the horse's ears with tick insecticide. Always check with your veterinarian first, however, before using a new product.

Despite horses' short and slick coats of hair, certain mites and lice find their way into it.

Lice will spend their lives leeching off the host equine. Both biting and sucking types of lice eat the animal's hair and skin, causing irritation and hair loss. Sucking types of lice live only a few days but if they heavily infest a horse, they can contribute to anemia. Unfortunately, lice are found in Florida nearly the year round, and usually on a horse's head, its mane and at the base of its tail. All lice can be transferred from animals in the same barn or pasture by flies, animal contact, contaminated tack and bedding.

Mites — about five different species — do their part to make Florida horses uncomfortable. Mites generally tunnel just beneath a horse's skin. As they drill in, they secrete a toxin that causes the horse's affected skin to flake off. Mites are the culprits that transmit contagious mange. Controlling external insects as much as possible is vital to Florida horse owners because these bugs often transmit illness among horses, and they infect animals with internal parasites such as bots. To minimize their effect takes some strategy and weapons, among them:

— *Wherever possible, move horses from low lying pastures into stables overnight.*

— *If the biting insect problem is severe, consider screening the stall.*

— *Keep the horse sprayed with repellents.* Currently high on the recommended list is Equicare Flysect™ Super-7, a pyrethrin product that has a seven-day residual effect. The Equi-Shield Fly Repellent Spray and Gel is being recommended by equine skin specialists for horses' ears and the base of the mane and tail. Many horse owners are reporting remarkable success with a half-and-half dilution of Avon's Skin-So-Soft applied as a daily wipe. For a short ride in mosquito country, human mosquito repellents temporarily work on horses, too.

— *Keep the horse's surroundings clean.* Dispose of anything that attracts flying insects such as animal droppings, rotting food, garbage, standing water.

— *Consider installing ceiling fans to keep the barn cool and to keep flying insects on the move.*

Principle Florida Equine Diseases

Nothing can ruin horse ownership like an unhealthy horse. And, despite their size, horses are surprisingly delicate. Many horse owners marvel at the thought of wild horses roaming free . . . risking colic, founder, worms, respiratory or viral infection, snake bite, foot problems, skin ailments, poisonous plants, tetanus. That early equines ever survived at all seems miraculous.

Yet many horse trainers and veterinarians say humans have created most problems for horses. To achieve size, color, beauty or speed, humans breed one problem horse to another. Thus, crooked limbs persist, bad temper is passed on in some breeding lines, and reproductive problems common in yet other bloodlines appear in successive generations. In the wild, horses that colicked easily or grew too lame to keep up with the herd, died. Thus, physically weak horses bore few young and had less opportunity to pass along bad traits. Humans not only have tampered with breeding, but also use horses in ways nature never intended — such as jumping seven-foot puissance walls. Moreover, many horses are kept 24 hours a day in unnatural habitats such as box stalls. The result is a whole new world of equine medicine that must constantly strive to help horses survive their artificial, sometimes hostile, environment.

In the 1800s, for example, many of the diseases of horses were chalked up to nervous disorder, brain inflammation or what was more generally called "mad staggers." Today, medicine has more clearly defined these perpetual diseases and found ways to treat them, even prevent them.

Equine Encephalitis (Sleeping Sickness)

Some veterinarians balk at the "Sleeping Sickness" nickname for Equine Encephalomyelitis, also called Encephalitis. The viral disease gained that pseudonym because affected horses appear dull and listless while standing and may walk in a staggering, drowsy fashion. Actually, Florida horse deaths due to the disease are reprehensible because available vaccinations offer protection. Yet in 1982, for example, 202 horses in the state died from the disease. Three types of Equine Encephalitis can invade Florida horses: Eastern Equine Encephalitis (EEE), Western Equine Encephalitis (WEE) and Venezuelan Equine Encephalitis (VEE). It is the first two strains that the annual vaccination guards against. VEE has not appeared in the United States since 1971. Still, some veterinarians recommend a VEE innoculation every three years.

Encephalitis occurs throughout Florida and is transmitted by mosquitoes and possibly other biting insects. Insects are thought to carry the disease from infected birds to horses. Once a horse contracts the disease, there is no guaranteed treatment. The animal's central nervous system is affected, so an owner may first notice the disease when the horse walks into objects, displays poor eyesight and/or runs a fever. Mild cases do recover, but mortality is 50 percent with WEE and 90 percent with EEE.

Officials warn Florida horse owners who live near boggy mosquito areas that their animals are at highest risk. Their horses should be sprayed each evening with insect repellent and the owners are encouraged to screen their horse stalls. Diagnosis for this disease should be made by a qualified veterinarian because symptoms can be similar to those of rabies, tetanus, colic or even some poisons.

Current vaccination procedures are two injections in early spring, given two to four weeks apart. Pregnant mares should be vaccinated approximately 30 days before foaling. Foals should be vaccinated at between eight and 12 weeks old. Some areas in the state occasionally step up vaccination programs because of unusually high outbreaks of this disease, so owners should stay up on the programs recommended for their locale.

Equine Infectious Anemia (Swamp Fever)

Equine Infectious Anemia is a virus that attacks the horse's red and white blood cells. Some horses contract so severe a case that they are diagnosed as acutely or chronically ill. Such victims often die within two weeks.

Symptoms of the illness are weakness, depression, rapid weight loss, anemia, and swelling in the limbs and lower body. Some chronic cases show signs of marked improvement before sinking once again into the symptoms and succumbing.

No vaccine is available for EIA. As for treatment of acute or chronic cases, little more can be done other than to make the affected animal as comfortable as possible.

Today the Coggins test is the standard diagnostic procedure. Developed by Dr. Leroy Coggins in 1970, the test analyzes blood samples and reports the findings as either a "positive" or "negative" result. This refers to whether the horse is carrying the EIA disease antibody (positive) or not carrying it (negative).

Another EIA test, the Competitive Enzyme-Linke Immunosorbent Assay, or CELISA test, affords a faster diagnosis. However, the new test is not used in Florida state testing laboratories because testing officials say it is more costly than the Coggins test. This fact, coupled with the same paperwork as the Coggins test, affords no great advantage to horse owners.

It is precisely because the Coggins test reports the presence of EIA antibodies that it has caused a storm of controversy among horse owners, because any positive reactor in Florida (and many other states) must be strictly quarantined or destroyed.

Up to that point, EIA is black-and-white clear. Clear, that is, but for the fact that another type of horse can earn a "positive" Coggins test result. This animal appears healthy, but its positive test result earns it an "inapparent carrier" status. Such carriers, according to law, also must be either destroyed or quarantined because current medical thinking is these horses act as EIA reservoirs, that biting insects such as horse flies and mosquitoes carry EIA-infected blood from carriers to healthy horses. Some "inapparent carriers" have been known to harbor EIA for more than 18 years, all the time remaining healthy themselves. Further muddying the swampy EIA waters is the fact that some veterinarians now are speaking up, contending that "inapparent carriers" do not transmit the disease, that it is the chronically or acutely ill animals that pose the real threat.

But the stringent EIA laws seem unlikely to soon change. And any horse owner who receives a positive Coggins report is expected to follow state law. That law says, in part, that current (less than a year old) Coggins test papers must be produced for a horse that is vanned to race tracks, rodeos, shows, boarding stables at fairs or other such assembly points, at public or private horse sales, and when a horse moves across state lines (to or from another state or country). Positive reactors also may not be bred. Thus, a healthy but "inapparent carrier" is useless except to a stay-at-home, backyard rider.

A quarantined horse must remain on home turf until it dies or is moved either to a research facility or to a slaughter house. Quarantine space is generally accepted by Florida officials to be 200 yards or more from any other horses.

As a result of these strict laws, some horse owners have refused to have their animals tested. They fear the test results. Veterinarians say, however, the chance of receiving a positive test result is more remote than that of a pastured horse being struck by lightning. Other owners who haven't had their horses tested or who have postive reactors have chosen to "fake" necessary test papers and contrive to travel with their horses to shows or get-togethers. These owners contend that no proof exists to show the "inapparent carrier" transmits EIA to other horses; that until there is such proof, it should be presumed carriers do not transmit.

Most veterinarians disagree. To them, the Coggins test remains a most reliable detection method and to place carriers among healthy horses, they counter, is reprehensible. Meanwhile, the state's Department of Agriculture has issued rule changes on displaying Coggins test papers. They must be either original copies or notarized photocopies. Tampering with test papers is a felony.

While research to find an EIA vaccine continues, state veterinarians have drawn up the following recommendations for Florida horse owners to deal with EIA:

- — Comply with the laws and regulations.
- — Test all horses in a band or herd for EIA.
- — Strictly isolate or destroy infected animals to prevent exposure of others horses or foals.
- — Bury deeply or burn carcasses of known infected animals.
- — After positive animals are eliminated, retest the remaining herd.
- — In high risk areas or herds exposed to other horses, test the herd at regular intervals.
- — Control biting flies, mosquitoes and lice.
- — Use only sterilized hypodermic needles and surgical instruments on each horse.
- — Avoid use of common horse equipment such as bits and grooming equipment.
- — Maintain 200 yards or more of separation from neighboring herds whose status is infected or unknown.
- — Isolate newcomers to the premises and retest in 60 to 90 days before commingling the horses.
- — Isolate and test any animals suspected of having EIA.

Equine Influenza

This infectious horse virus prompts loss of appetite, depression, slight nasal discharge, fever and cough. It often pops up at any stressful, crowded location such as at races, shows or sales. As with other equine diseases, it is better prevented than later treated, although treatment often succeeds.

Affected horses should be rested. Veterinarians sometimes prescribe antibiotics or corticosteroids.

Prevention is by vaccination but may be given only to horses who will be at risk of exposure at race tracks or showgrounds. Foals often are vaccinated at age two to four months. Although reactions to the vaccination are rare, some horses do form a "goose egg" after their flu shot. This bump at the shot site disappears in about a week, but should be watched for secondary infection.

Equine Viral Arteritis

Often dubbed EVA by veterinarians, this virus attacks the horse's upper respiratory tract. Affected horses are feverish, their eyelids and legs may be swollen and the animals suffer from chest inflammation and a runny nose. Male horses' genitals sometimes swell. Most serious, however, is the disease's effect on pregnant mares; they may abort their foals with the onset of the virus symptoms. Medical studies suggest 5 to 16 percent of any given horse population harbors EVA. Although some horse books state there is no vaccine for EVA, a live virus vaccine, used with restriction, is available in states with important Thoroughbred industries. That includes Florida. Vaccinated horses may not be bred, hauled, shown or raced for 28 days after they are given the shot.

Florida veterinarians do not advise Florida owners to arbitrarily vaccinate their horses for EVA, but barns with high horse traffic arriving for sales, shows or breeding — especially from out-of-state — are advised to quarantine new arrivals as a precaution.

Mortality rate for EVA is low among adult horses. Heaviest losses come from aborted fetuses at a rate that can reach 80 percent.

Equine Rhinopneumonitis (Equine Herpesvirus)

A mouthful to say, this virus' name is shortened to "Rhino" by most horse folks. The less eloquent among the equine set have dubbed the disease "the snots," a more or less self explanatory term.

Death seldom occurs from this disease, but it can be dangerous because it causes complications such as pneumonia, especially in young horses. In breeding businesses, Rhino assumes greater importance because it can cause abortions. Timing of such abortions is one indication of whether a horse suffers from Rhino or Arteritis. Rhino's abortions usually occur after the symptoms, when mares reach their eighth to eleventh month of gestation; arteritis, on the other hand, usually induces rapid abortion as symptoms appear. Two positive aspects of Rhino offer horse owner's some hope. One, horses usually gain some natural immunity once Rhino has run its course, although that immunity last only a few months. Owners, especially if planning to breed their stock, should take advantage of the second good aspect, vaccination. Pregnant mares usually are vaccinated at gestation intervals of five, seven and nine months.

Equine Piroplasmosis (Tick Fever or Bilary Fever)

Piroplasmosis, like Equine Infectious Anemia, is transmitted by an insect. Florida's tropical horse tick transmits a protozoan parasite that attacks a horse's red blood cells. The result is fever and anemia. Colic and, later, pneumonia, also may occur. Only an estimated 10 percent of affected horses die. But as in EIA, horses infected with Piroplasmosis may remain carriers of the disease. Thus a single female tropical tick feeds on a carrier horse, produces 2,000 to 3,000 offspring which, in turn, transmit the disease to other healthy horses.

Florida veterinarians recommend that horse owners watch for ticks — primarily in horses' ears and false nostrils (skin flaps near the true nostrils). Advice on tick sprays are available thorough veterinarians or local County Extension Services. Care must be taken with this disease because it is serious enough to warrant quarantine. Infected premises may be placed under quarantine until no horse is found infected for six months and no tropical horse ticks are found during that same six-month time. Horses may be moved to another premise, but those premises then would be under quarantine. State officials say affected animals may leave quarantine three days after acceptable treatment by a veterinarian and providing that USDA and Florida Department of Agriculture supervised tick control measures have been followed.

Strangles

Strangles is an equine bacterial disease in which, as the name indicates, the throat and upper respiratory tract of the horse are infected.

A thick, whitish nasal discharge and the horse's refusal to eat or drink are an owner's first hints of the ailment. A healthy horse usually overcomes strangles in about three weeks. Occasional complications of strangles occur if infection spreads to other organs that may abscess.

Strangles is highly contagious. Animals at sales, shows, and boarding barns are most susceptible if one animal among them catches strangles. The disease is spread by droplets of infected

The horse in distress. An ill horse often exhibits physical "body language" symptoms such as rolling (top), lying down and biting at its side (center), or standing with head hanging in the classic "sick horse" posture (bottom).

A vaccine available from veterinarians eases the pain of horses suffering from either laminitis (founder) or colic. The vaccine does not prevent either ailment, but does block the horse's production of endotoxins, which cause the pain and side effects associated with colic or laminitis. Called a "laminitis shot," it is given in 2 injections, 2 weeks apart, and followed up each year with a booster shot.

---

material that is inhaled or eaten by healthy horses. Affected horses should be given immediate veterinary care and isolated. Following recuperation, the infected stall should be thoroughly disinfected. Strangles bacteria reportedly has been found to live in empty stalls for up to six weeks. Vaccination is available.

Colic

A book could be written on just the subject of equine abdominal pain that is generally termed colic. Such a book, however, would still not answer the horse owner's primary question: Why does it occur?

In fact, there is some disagreement as to whether colic ought to be classified as a disease or as a symptom of many other diseases. For nearly any disruption in a horse's normal health can contribute to a colic bout. To make matters more confusing, colic symptoms can be mistaken for other medical problems such as hepetitis, peritonitis and tetanus.

Definitely known about colic is that it is the leading killer of horses everywhere, that in its acute form it is exceedingly painful, and that it can be caused by many factors such as gorging on cold water when the animal is overheated, lack of water, overeating of feed, accidental twisting of the intestines, and excess gas due to feed that is either moldy or too rich. Recent studies have found one major cause of colic to be the large strongyle worm whose larvae can severely damage the intestinal system. Furthermore, if worm-weakened intestines are further stressed by an intestinal blockage they can burst, sending fecal material into the horse's abdominal cavity. The usual result is death.

In Florida, sand is another major contributor to colic. Horses unintentionally consume sand while grazing. The sand sits in the intestinal system, and not moving through, hardens to a cementlike consistency. A bout of colic takes on serious dimensions because horses have what is termed a "one-way" digestive system. Muscles do not bring food back up "from whence it came." If feed is toxic or disagrees with the horse, the animal is stuck with it and colic is a predictable outcome.

Predictable to any horse owner who does his or her colic homework are its symptoms: patchy sweat, heavy breathing, depressed behavior, standing "parked out," numerous attempts to urinate or defecate (often with no success), nipping or kicking at sides, frequently lying down and rolling. Less common, more desperate symptoms include dehydration, pale or bluish mucous membranes and rapid heartbeat (possible onset of shock). Once the horse owner spots any of these symptoms, a call to the veterinarian is in order. Some veterinarians put the owner right to work taking vital signs such as temperature and respiration. At one time, it went without saying that a colicking horse should be walked to keep it from rolling. But some veterinarians believe many nervous owners overdo the walking. Light walking, even permitting the horse to lie down to rest, may be okayed by some veterinarians as long as the horse does not roll violently. Unfortunately, colic prevention is not always easy. But like so many equine ailments, it is usually more easily prevented than treated after it has a hoof-hold. Prevention is easier, too, on the owner's bank account. Some veterinarians recommend periodic mineral oil drenches to help keep the "pipes" clean, particularly of sand. Other practitioners contend that an average horse owner cannot get enough oil down a horse without a stomach tube to make much difference. Still, if owners recognize a few equine behavior and digestive basics, colic bouts can be held to a minimum.

First basic: Nature designed horses to eat virtually all the time. Stalled horses usually eat hungrily just twice a day. This works against the horse's natural digestion. If a horse must remain stalled, try to feed it at least three times a day, preferably four. Make sure it has plenty of hay. Poor food digestion contributes to colic, so have the horse's teeth checked by either an equine dentist or a veterinarian once a year.

Second basic: Horses are creatures of habit and respond best to scheduled work and feed routine. Horses have amazing inner clocks that signal when it is time for each activity. Change in routine now is recognized as an equine stress, and stress is a major cause of colic.

Third basic: Horses have a limited palate. They do not eat meat or other exotic treats. Their digestive systems do not cater to moldy feed or hay. If a horse eats its food too quickly, it may be famished — or that may just be that animal's greedy personality. A handful of fist-sized river rocks in a feeder may slow down the wolfer.

Fourth basic: Never give unlimited water to a sweating, overheated horse. Similarly, don't work a horse immediately after it has been fed. Do not feed cattle feed to horses.

Fifth basic: Idle horses may be more prone to colic than active, fit animals. If your horse is boarded and cannot be ridden everyday, try to ensure the animal will be turned out to pasture or exercised daily by a friend or handler.

Sixth basic: An estimated 80 percent of colic cases are related to parasite, primarily strongyle, infestation. Florida horses must be dewormed regularly.

Seventh basic: When a horse colics, seek immediate medical help. Only skilled handlers can differentiate between truly serious colics and minor bellyaches. Home colic remedies are not wise treatment because some drugs can impede digestion and make colic worse. The ultimate outcome of many colic cases is surgery. Only a few excellent equine centers are equipped to provide equine surgery. One is the University of Florida in Gainesville, where many Florida horses are taken for surgery at the College of Veterinary Medicine. College officials say about 350 colic cases are admitted each year — one of the highest such caseloads in the nation. They attribute the high numbers to the state's large horse population, the sandy soil and the mild winters that benefit parasites.

University of Florida medical personnel estimate an 80 percent colic success rate. Numbers would be higher, officials say, if owners were quicker about seeking treatment. The remaining 20 percent are lost to a variety of complications.

Tetanus (Lockjaw)

Any horse that has not been vaccinated for tetanus toxoid is at risk for contracting a horrible disease commonly called lockjaw. The ailment earned its name from the symptom that prevents an affected animal from swallowing. Part of the disease keeps the animal stiff, while other symptoms cause some muscles to contract in spasms.

Tetanus is caused by bacteria in "things you find in the barnyard every day." The bacteria infect any puncture wound, even a surgical opening, and thrive in abscesses or under scabs. Unvaccinated horses suffer about an 80 percent mortality rate if infected by tetanus. For unprotected horses, treatment generally is of little help, but some recoveries occur with use of mega-doses of tetanus antitoxin and antibiotics.

Prevention is the key. State veterinarians recommend that all Florida pleasure horses be vaccinated annually for tetanus. Broodmares may be vaccinated one month before foaling, and foals should receive the antitoxin at birth and again when three months old. In case of injury, horses usually are given a booster shot.

Rabies

Horses get rabies, but less than 60 horses in the nation contract the disease each year. There is no treatment for horses who get rabies, so some veterinarians advise horse rabies vaccinations — especially if there is an outbreak among nearby domestic, farm or wild animals.

Anhydrosis (Non-sweating Syndrome, Dry Coat)

A sharp-eyed horse owner notices a subtle change in his or her mount following a summer late morning workout: There is no lathery sweat on the horse's body. In addition, the horse seems unusually short of breath. A few weeks later the animal's facial hair is nearly gone.

The diagnosis may be anhidrosis. The horse is not properly sweating. A dilemma faces the owner: Move the horse to a cool (maybe even air-conditioned) stall for the summer; move the animal to a cooler climate where the ailment usually corrects itself; destroy the horse or subject it to treatment that may be expensive but not particularly helpful. Failure of horses to sweat was first reported by the British when they took their Thoroughbreds to tropical colonies such as India. Today, Florida veterinarians report an increase of non-sweating horses. And, whereas Thoroughbreds once were most affected, now every breed has individuals with the ailment.

Sweating is crucial in a warm, humid state like Florida. Horses, like humans, must sweat to regulate their body temperatures. A horse whose body cannot cool itself down soon would die in subtropical heat. Thus, non-sweating horses are of little pleasure or economic value for any summertime activity in Florida. A few owners say they can put up with the syndrome by riding in early morning or late evening hours, but use of their horses is severely curtailed.

A variety of treatments are attempted to correct non-sweating syndrome, but none has been totally successful. Finding just the right one for each animal is a trial-and-error game.

The most common treatment is giving the horse oral electrolytes, which can be found in some horse supplements, or are available from veterinarians. Success for this treatment ranks about 50-50.

Another partially successful treatment calls for thyroid supplements. Current thinking is that anhidrosis is caused by more than one physical abnormality. Whatever the causes, owners who suspect their horse has this syndrome should make the animal comfortable immediately and call in a veterinarian.

Skin Ailments

As mentioned earlier, a great many bugs bug Florida horses. Mites, ticks, lice, flies, pinworms all take their toll on a horse's hide, causing welts, hives, bloody bumps, infection and hair loss as the horse rubs the affected areas.

Mites, particularly, can cause different tongue-twisting types of mange, usually the psoroptic, chorioptic and sarcoptic types in Florida. Psoroptic mange sprouts up under the horse's mane, the base of the tail and on the inside of the legs; chorioptic mange is found on the lower legs; sarcoptic mange is commonly on the horse's head, neck and shoulders. To the average Florida horse owner, however, the technical names may mean little; mange is mange and any way it is spelled it takes up extra hours of a horse owner's schedule to get rid of it.

Infected animals usually reveal their mange with some zeal. They rub against trees, buildings, fence posts and other horses in an effort to ease the itching. Poorly nourished, rarely groomed animals are much more prone to mange, so the ailment signals that more care of the animal is needed.

Florida's high yearly rainfall and humidity keeps some mange

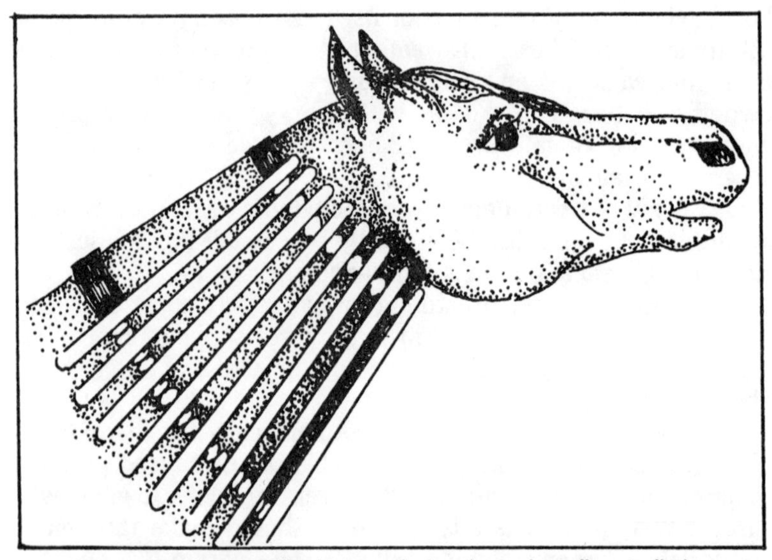

What looks to be a medieval torture device, the neck cradle actually is a useful apparatus that keeps a horse from chewing on a healing wound or ripping off bandages.

unmanageable as mites can live in moist stall woodwork or in damp bedding and will reinfect the horse.

Insecticides are useful in treating mange. They should be administered according to a veterinarian's direction. Ringworm, which is a fungus, also can make a horse look downright moth-eaten. Like mange, ringworm is passed along easily. Many farm animals — dogs, cats, mice — harbor the fungus and give it not only to the horse but to its owner. And like mange, ringworm can travel from horse to horse on shared grooming equipment and tack.

Ringworm can be treated with any good fungicide available at a farm supplier or saddlery.

It is important that such ailments be diagnosed by a veterinarian because allergies sometimes can mimic the symptoms of parasites. In such cases, the treatment would, of course, be different.

Some horse owners become concerned when they see dandruff in the mane or tail of their steed. Small amounts of dandruff are normal. If the hair falls out to any extent, however, it could mean a problem that should be seen by a medical person.

A lady's brassiere makes an economical, yet workable, eye-patch for a horse that must have one eye covered after an injury or surgery. Cut out one side of the front for the good eye. Adustable straps go over the horse's ears, the back snaps behind the horse's jaw.

A common Florida skin problem is generically termed "rain rot." It's an eruption of the skin with some hair loss and becomes bothersome during long-term wet seasons. It usually disappears with a few baths and reappearance of sunny days.

This condition may be related to what the British call Mud Fever, a skin irritation caused by too much moisture, especially where mud dries on the horse, such as on the lower legs. It is common in Florida on thin-skinned horses and is relieved by carefully washing and drying the affected area. In dry weather the condition usually abates or disappears completely.

Horse owners, especially those new to Florida, must take extra precautions with any area on their horse that has been rubbed raw. Simple saddle sores or scrapes can turn into full-fledged infections or into summer sores, sometimes called "swamp cancer" or "leeches" by native horse owners. Although a fungus can cause summer sorelike patches, the genuine summer sore is actually a fibrous tumor called Bursatti in most veterinary texts. The sore can be caused entirely (or aggravated) by larvae of the *Habronema* stomach worm that migrate from flies' bodies to a pre-existing wound. Summer sores invariably are situated in slow-to-heal areas on the horse such as the knee, or occur in areas the horse can keep inflamed by chewing or rubbing. Many a veteran northern horse owner new to Florida has been frustrated to find that no sooner has the summer sore scab disappeared then a new sore appears on the same site. For this reason, many veterinarians recommend cauterizing summer sore sites with a sterilizing hot iron. Prevention of summer sores is rooted in basic hygiene. Any neglected scrape that is buzzing with flies begs to become a summer sore. Ointments are available that repel flies while keeping the wound dirt-free. Stalls should be kept clean and as fly-free as possible. Horses should be dewormed regularly to discourage stomach worms.

Saddle sores can also be a horse owner's headache. Badly fitting saddles, or saddles used without pads, can rub dirt into the horse's back area. The hair is rubbed off; dirt and sweat are ground into the area. Hot, swollen spots erupt. If left unattended, they can become infected or may develop abscesses. Severe saddle sores that heal and are irritated repeatedly eventually lose the hair completely and grow callous-hard.

Preventing saddle sores is easy if owners properly fit horses' saddles and use clean, good fitting saddle pads. An extra precaution when readying to ride: Set the saddle and pad on the horse's back nearer than usual the animal's head, then slide the pad and saddle back to the desired location. This smooths the lie of the hair under the saddle, reducing the risk of saddle sores.

Moving the saddle and pad back into proper position after it is on the horse smooths hair down and helps prevent saddle sores.

Equine skin ailments often erupt from a dietary cause. Hives as large as dinner plates can pop up because the horse gobbled a good-tasting pasture plant that produced an allergic reaction. A rashlike profusion of small hivelike bumps along the stomach can signal urticaria, a possible upset stomach. Only the veterinarian, however, is equipped to make proper diagnosis as to whether a bumpy horse got its welts from "without" (flies, etc.) or from "within" due to a food allergy.

Florida's summer sun — often indistinguishable from Florida's winter sun — can burn a horse's hide as surely as it burns human skin. Especially prone to sunburn are light-colored horses such as Palominos, duns and white horses. No area on the horse is as vulnerable as the nose, around the eyes and near the tail. Just as with human sunburn, the horse's skin gets red and swollen, sore to the touch, and peels. Some horse owners have reported success with human sunscreen preparations. Likewise, some horse fly-sprays now contain sunscreen. Ointments such as Corona or bag balm also aid in preventing sunburned skin.

Florida Horses' Foot Problems

Doubtless every horse owner soon learns that old horse world cliché, "No foot, no horse." Despite its hackneyed use, the reportedly 5,000-year-old phrase is virtually horsedom's universal truth.

Shoeing

Do Florida horses need to be shod? One of the state's most respected horseshoers says he has "never met a horse that didn't need shoes." But then, all dairy farmers say everybody needs to drink milk, too.

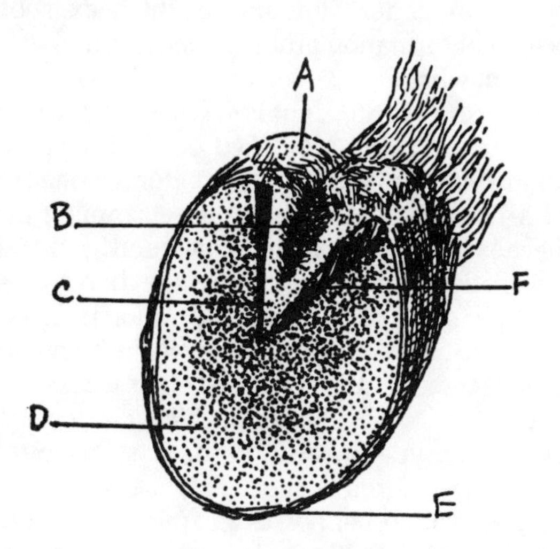

THE HOOF: A. Heel; B. Cleft of frog; C. Frog; D. Sole; E. Toe; F. Bar (one of either side of frog).

Some horseshoers say that because Florida's soft sandy or grassy turf doesn't naturally wear down the horse's foot, it needs to be shod. Other farriers and some veterinarians say that no shoes are necessary for the same reasons.

Whether to shoe a horse is a decision horse owners must make themselves. It should be based on several considerations. First is the condition of the horse's foot. Badly cracked or deformed feet can be improved with shoes. At the same time, the horse owner must realize that horseshoe nails can damage (minimally) the hoof, too.

A second consideration is the horse's use. A horse that is used often on hard roads needs shoes. Then there are horse shows with rules requiring show entrants be shod, particularly park-type breeds such as Saddlebreds and Tennessee Walking Horses that wear built-up shoes. Of course, dressage competitors generally have hooves that are trimmed to a long toe and low heel, before shoeing, to promote a longer stride. Conversely, a rounded toe on a more upright-standing hoof, before shoeing, affords a shorter stride and is seen more often on, say, a Western pleasure horse.

Another consideration is how the horse travels. Most horse shoers agree that after a horse matures, usually at four or five years, corrective shoes afford little help — and can even promote other problems in higher joints such as the hock, shoulder or hip. Young horses' conformation problems, such as toeing in or out, can be altered somewhat.

How the horse is worked, not how it is shod, affects the animal's movement. Farriers often are called out to correct forging problems, meaning when a horse hits the bottom of one front foot with the toe of a hind foot. New shoes or retrimming may not help, though, because horses forge from fatigue that is often coupled with conformation faults. Presuming the horse's needs haven't prescribed shoes so far, the third consideration is an owner's wallet. A shod horse must have its shoes removed, its hooves trimmed and the shoes replaced about every six to eight weeks. This "reset" is cheaper than the initial shoeing, which can cost $20 and up, depending on the type of shoes used. Most horse owners opt to have just the horses' forefeet shod. This is a good idea if the horse is turned outside with other horses and might kick a pasture buddy. Regardless of whether the horse is shod or not, its hooves must be trimmed regularly. Trimming every six to eight weeks assures the horse is standing evenly and balanced on all four feet. Such care also catches problems such as sand cracks before they grow severe. Trimming costs about half of shoeing.

Horseshoers and most veterinarians can tell a great deal by looking at a horse's hoof, for the hoof often indicates past health problems. Such ailments such as laminitis (founder) and navicular disease often leave tell-tale ridges, either raised or sunken, on the hoof.

Lameness

A horse's foot problems range from simple abscesses to

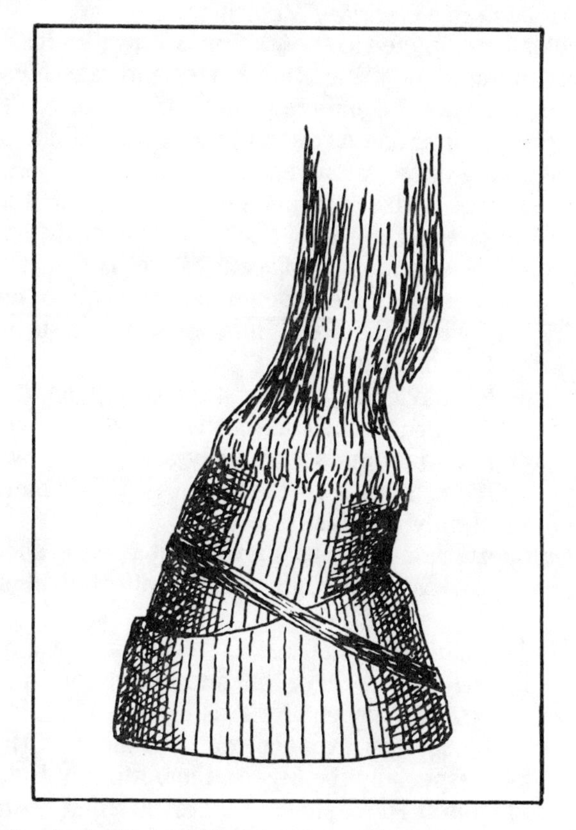

A built-up hoof. Added weights such as the one pictured are seen on gaited horses, usually American Saddlebreds or Tennessee Walking Horses, and promote high-stepping leg action.

complex ailments such as laminitis (also called founder) and navicular disease.

It's no wonder, then, that horse owners tend to panic when an animal goes lame. Diagnosing a cause of lameness can be maddening. A stiff shoulder, for example, can be stiff from work, can signal a touch of arthritis, or can be the initial harbinger of navicular disease.

The insidious navicular disease is, of course, one horse problem that is prevalent throughout the U.S. Because Florida is one of the nation's biggest horse states, veterinarians treat scores of cases annually. In fact, navicular has been called "the most common cause of lameness in show horses" by doctors at the University of Florida's College of Veterinary Medicine.

Navicular is not limited to show animals, however. It also occurs in working animals such as cutting horses and race horses.

One peculiarity of the ailment is that it is difficult to detect. An affected horse is lame one day, sound the next. Definite diagnosis is made by radiography (X-Ray) and by blocking the horse's lower leg nerves with pain killer to isolate exactly what spot is sore.

Soreness is deep in the hoof itself, in the sensitive navicular bone. Pain arises when joints adjacent to the navicular bone are inflamed. Bone and cartilage degenerate, causing foot tendons to stop gliding easily over the bone. Bone spurs and cysts may form; ligaments and tendons can harden.

Treatment for navicular disease traditionally has been with anti-inflammatory drugs such as cortisone or phenylbutazone (Bute). Persistent symptoms occasionally are treated by severing the pain-conducting nerves in the foot. All these methods are drastic and used only as a last resort.

More modern research has shown that special shoeing can manage navicular cases enough to allow affected horses to be ridden.

Nobody is quite sure what causes navicular. But many veterinarians and farriers now believe that improper hoof trimming and shoeing can aggravate navicular disease.

A healthy horse foot expands and contracts as the animal walks. The foot, especially the frog on the bottom of the foot, acts as a vital blood pump. Horse shoes that are too small or improperly shaped can cause contracted heels, that is, a hoof that is too narrow at the heel. Just as human feet don't operate properly if crammed in too-small shoes, so the horse's hoof does not pump properly if it is constricted.

Not only proper trimming and good-fitting shoes benefit the navicular horse; so does exercise. Only by moving around does a horse's foot properly circulate blood.

The very fact that the horse's hoof is a major blood pump may make it more prone to laminitis, also called founder. This painful condition is a congestion of the hoof's blood supply as the volume of blood (and pressure) inside the foot increases.

A horse owner usually cannot mistake severe laminitis. Because the ailment usually affects one or both front feet (rarely rear feet), the horse props itself on its hind feet. The owner may think at first that the horse is paralyzed because it refuses to move. If pain is particularly severe the horse may lie down and the owner mistakes the founder for colic. A veterinarian should be summoned immediately. Until the doctor arrives, an owner can give some relief by spraying or soaking the horse's feet with cold water.

Laminitis, like so many other horse ailments, often can be prevented. The usual cause is a sudden dietary change — such as when a horse breaks into the feed bin and eats too much food. Similarly, occasional laminitis occurs if horses are turned out in a newly opened, lush pasture.

Other causes of laminitis include too little exercise, long workouts on a hard surface such as concrete or asphalt, uterine infections and pneumonia.

Corns also can prompt a bout of laminitis. Corns are closely related to, even caused by, bruises. Purplish-colored smudges on the bottom of the horse's foot, corns are known to abscess, affecting tissues and bones inside the horse's foot. When the pressure builds in the foot, extreme pain and lameness result. Seeking relief, the horse may lie down. Infected material may erupt out the top of the hoof, which should be treated by a farrier or a veterinarian.

Two common hoof ailments that occur in Florida are thrush and cracked hooves.

Neglectful owners may find their horses' feet are actually rotting away with thrush. Swampy, mucky pastures or urine-sodden stalls harbor bacteria. The condition is easily recognized as the frog of the hoof sloughs off and produces a putrid odor. If further neglected, thrush can cause permanent lameness.

Treat thrush with any of the products on the market, or one part

"Too few veterinarians do foot work. And horse owners don't know enough about what horseshoers are doing to their horses." Ron Dyer, Florida horseshoer

---

liquid laundry bleach mixed in two parts water can be applied. Do not mix commercial thrush preparations with bleach. It is important to clean a horse's hooves daily, and absolutely crucial to clean a thrush-affected foot that often because thrush bacteria live only where the air has been cut off. Thrush is found most often in horses' hind hooves because more owners fail to clean those feet.

Cracked hooves are another problem. Contrasting with thrush's mushy-moist environment are the dry, sandy conditions that crack hooves. Quarter cracks crawl up the sides, or quarters, of the hooves; sand cracks may appear anywhere, but often start at the top of the hoof and split downward. Both types of cracks cause grit to enter the hoof's tissues and become infected.

Some horses have crack-prone hooves that farriers and veterinarians cannot explain. The theory that the hoof on a white-haired leg is more fragile has not been proven. Bad nutrition may contribute to cracks but the majority of horses will never have major cracks if a hoof dressing is applied to the foot two or three times a week. Many good dressings are available and help seal in moisture, which keeps equine feet pliable and easily doing their jobs of carrying weight and pumping blood. Caution should be used with hoof dressings, because what seals in moisture also can block out moisture. Heavy waterproof sealants such as pine tar can damage hooves if overused.

Severely dry hooves may take several weeks of daily care to correct. Soak the hooves in water, pat them dry with a towel and coat the feet lightly with a lanolin-based dressing. If the horse is kept in a clean dry stall, add a light touch of dressing to the top of the frog. Owners should remember, too, that stall bedding such as shavings can pull moisture from the horse's feet.

With some luck, cracks and other foot problems may become extinct. New synthetic products are in the works not only to repair serious cracks, but to glue on horseshoes instead of nailing them on.

THE POISONOUS SIDE OF FLORIDA

Florida beaches are the state's main attraction, but an increasing share of the nation's nearly 10 million equines and their owners are lured by the Sunshine State's balmy breezes, its lush trails and year-round pastures.

Those very trails and pastures, however, harbor subtle dangers to Florida's horses in the way of poisoning. More than a dozen varieties of toxic trees and weedy shrubs are common to the state, grow in fields and may end up in bales of hay.

Some poisonous plants are repulsive weeds, even to horses, and therefore seldom get eaten. Others, however, are alluring, even attractive enough to be sold commercially as ground covers or shrubs.

The following list includes many of the more commonplace and potentially hazardous plants in Florida:

Bitterweed — a flowering plant that, when blooming, ranges in height from six inches to three feet. Leaves are about an inch long, very narrow and numerous on stems and branches. Flowers look like daisies or black-eyed susans but have bright yellow petals and centers. Found primarily in central and north Florida and across the Panhandle to Alabama, the plant is seen along roadsides, near dumping areas and in fields. Because the weed favors bright sunlight and sandy or clay soils, it is seldom found in swamp areas. All parts of the plants are poisonous, but reported cases are generally light, producing only mild discomfort. As the weed's name indicates, it tastes bitter; the result is many horses do not bother it. Poisoning symptoms: abdominal pain, salivating, occasional green nasal discharge.

Boxwood — this ornamental shrub is found in many household landscapes that horses or foals may have access to. The shrub is rarely more than five feet high, with bright green, shiny, oval leaves. Both the bark and the leaves contain a poisonous alkaloid substance. Fortunately, these plants have a bitter taste that most horses avoid. Poisoning symptoms: diarrhea, intense abdominal pain, possible convulsions.

Bracken fern — a coarse fern with long, sturdy stems and rootstocks that may be 10 feet long. The leaves and stalks are one to three feet long and the triangular leaf blades are one to three feet across. Bracken is found in open, sandy areas, open woods,

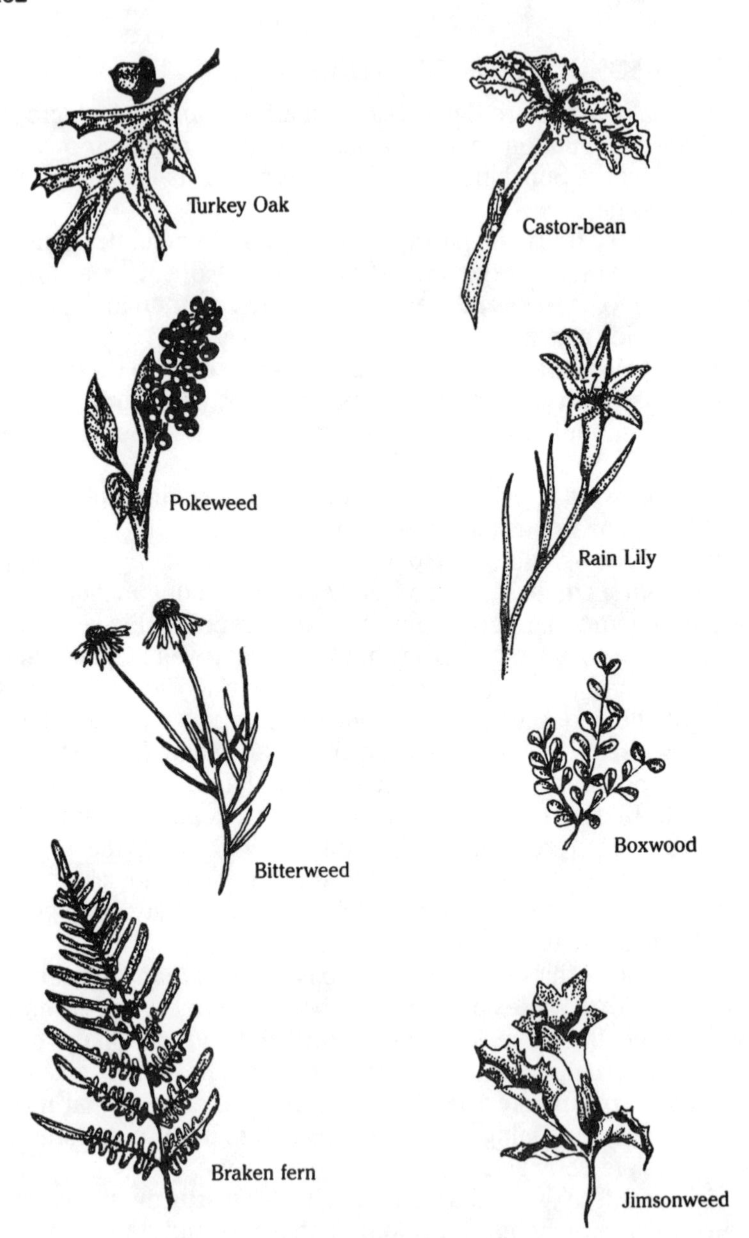

Poisonous Plants

pastures, and fields throughout Florida from Lake Okeechobee north. This fern may be inadvertently mixed with baled hay or bedding materials. Dried fern material also is toxic and tends to have an accumulating action, with most damage occurring after several meals that include the plant material. Illness occurs three or four weeks later. Poisoning symptoms: drowsiness, difficulty in swallowing. The animal may push its head against solid objects. From one week to 20 days after eating fern material, the horse grows weaker and can no longer stand. Some animals do rise and appear somewhat improved, but death usually occurs within several days. Hay should be examined for the dried bracken fern; if it is found, the bale should not be fed.

Carolina-jessamine (Evening Trumpet-flower) — one of the showiest but most harmful vines often used in Florida landscaping. The plant's main stems are gray and an inch or thicker. Oval-shaped leaves are one-half inch to two and one-half inches long and dark green. Flowers are fragrant and trumpet-shaped with five petals and are up to one and one half inches long. Seed pods are brown, flat and contain several small winged seeds. While the vine is mostly found in north Florida, it can grow as far south as the Kissimmee area. A high climber, it grows up small trees but also is known to trail along the ground. The plant's flowers, leaves and roots contain poison that chiefly paralyze an animal's brain and spinal cord motor nerve endings, resulting in an end to breathing. Many poisonings of livestock by this plant occur in winter when pasture grass may be poor. Pastured animals that have eaten these plants are not usually known to be sick until they are down. Poisoning symptoms: staggering, dilated pupils, feeble pulse, breathing difficulty.

Castor bean (Castor oil plant) — a shrub in central and northern Florida — a tree in south Florida — with long stems four to 10 feet high that are green, red or purple and sometimes covered with a waxy coating. Leaves, four to 30 inches across and held by stiff stalks, are star-shaped with five to nine "points" and finely-toothed edges. Flowers are narrow clusters six to 12 inches long and are creamy green or reddish brown colored. Fruit is oval, green or red and covered with stiff spines. This plant also has been used as an ornamental. It is found throughout the state, but thrives in rich soil, around dumps, and around Lake Okeechobee. All parts of the plant are poisonous, but the fruit (which contains the seeds) is the most toxic. It contains ricin, a poisonous protein. Poisoning symptoms: weakness, muscle tremors, emaciation.

Cherry laurel — a shrub or small tree that may grow 25 feet high. This tree is commonly used around homes in Florida as a landscaping shrub or tree. It is also often found in woods and hammocks and along fenced areas. The trunk is dark, crooked and may bear many lateral branches. Leaves are two to four inches long, oval-shaped, pointed at both ends and very glossy on the upper side with a few sharp points along the leaf edge. Flowers, pinkish white with five small, round petals, are one to two inches long. Its fruit is bluish-black, round and about a half-inch in diameter, borne in clusters of two to five. Leaves, twigs and fruit kernels smell peachy when crushed. The toxin in cherry laurel is hydrocyanic acid. It can cause death instantly if enough of the plant is eaten. Stricken animals usually linger with convulsions, depression, speeded respiration that grows weak and irregular, pupil dilation, eyes that stare, and a nose and mouth filled with foam. Respiratory paralysis is possible.

Crotalaria — a plant that ranges from three to six feet tall and is found in virtually every part of Florida. Its somewhat large, waxy leaves are four to seven inches long, dark green on their topside, lighter on the underside. Leaves may have a bristle at the tip. A spike of yellow flowers grows up on a tall stalk, resembling a snapdragon bloom. These plants are used as a cover crop in Florida to enrich the soil and to combat root nematodes. Just as often, however, the plant is found around buildings, along roadways and in abandoned fields. Although the leaves, stems, roots and seeds all have been found to contain a toxic substance, the podlike seeds, which are a quarter-inch long, have the highest concentration. This plant can be especially sinister because horses have been known to become affected by crotalaria poisoning as long as nine months after contact. At the same time, some farmers steadfastly maintain their horses have eaten these plants with no adverse effect. Poisoning symptoms: depression, loss of appetite, drooling, nasal discharge. Death can occur within five to 10 days, usually due to heart failure.

Hydrangea — this wild shrub may grow into a small 15-foot tree. It has large leaves, six inches or more long. Leaves, deeply lobed like oak leaves, are dark green on top and grayish, fuzzy beneath. Flowers are small, white and form a pyramid of heavily clustered blooms that may be a foot long. Old flower clusters turn purple-brown. Hydrangea grows near water — on river bluffs, at the edges of sinkholes and, from Leon County westward, among rocky outcrops. It shuns the sunlight, thriving best in the shade. Because it is an attractive plant, people have used it as an ornamental and thus spread the plant beyond its normal range. The wild varieties of hydrangea — as well as the hydrangea that is sold as an ornamental — contains hydrocyanic acid, a form of cyanide. Poisoning symptoms: abdominal pain, severe diarrhea.

Jimson weed (Thorn apple) — a weed that stands three to five feet tall with a main stem and branches that are smooth green, or sometimes purplish. Large, sharply pointed leaf blades, three to eight inches long, are thin and pointed at both ends. Flowers are single, funnel-shaped and fan out into a "star" that is usually white or purple-blue. Fruit is a pod, about one inch long, covered with hard, sharp spines. At least three toxic alkaloids are found in every part of the plant, but the seed pods are most harmful. Special care must be taken to watch for this plant being inadvertently baled in hay (the plant is just as poisonous when it is dried as when growing). To most animals the weed has a strong taste and therefore they usually do not eat it. Poisoning symptoms: rapid pulse and respiration, partial blindness, diarrhea, pupil dilation.

Lantana — another of Florida's native shrubs that is increasingly used as an ornamental for hanging baskets and landscaping. The plant reaches three to five feet in height with stems that may have sharp spines. Oval-shaped leaves, pointed at one end, are one to three inches long and scalloped along the edges. Flowers are delicate shades of white, yellow, pink, orange or scarlet and clustered in miniature "bouquets." Lantana grows wild nearly everywhere in Florida, primarily along roadsides and in fields. Poisoning symptoms: overall weakness, refusal to eat, mild diarrhea. Both the mucous membranes and the white areas of the animal's skin may show a yellowish color. In some cases, the animal suffers paralysis of the legs, and its skin cracks and peels, leaving raw exposed areas. Skin around the nose and mouth also may grow orange-colored, followed by cracking and bleeding that leaves the animal prone to secondary skin infection.

Oak tree — found virtually everywhere in Florida, the state's 28 species or varieties of oaks include turkey (also called black jack), scrub, white, red, laurel, water, live and swamp. All are members of the genus Quercus that contains a toxin authorities have not definitely identified. Some references attribute oak poisonings to tannic acid, while others explicitly point out that tannic acid is not the toxin. Regardless, oak acorns particularly can kill horses. Poisoning symptoms: constipation followed by diarrhea, loss of appetite, dry and cracking skin around the muzzle, increased thirst, weak pulse, depression and abdominal pain. Death of affected horses can occur within two days to two weeks.

Oleander — this pretty, willowy ornamental has been extensively planted in beautification projects. It grows fast throughout the state. The plant is classified as both a shrub and a tree, ranging from five to 25 feet tall. Older, larger trees have gray trunks with rough pores. Leaves are three to 10 inches long and smooth. Colorful flowers of white, pink, pale yellow, rosy pink or deep red are often trumpet-shaped with five petals about one-inch long. Some cultivated double varieties contain many petals. Oleander contains two toxic substances that resemble the digitalis glycosides. Poisoning symptoms: rapid pulse, weakness, profuse sweating, severe abdominal pain. Less than one ounce can kill a horse, so instances in which a horse has consumed much of the plant harbor little chance of survival. Note: This plant also is very poisonous to humans. One leaf can kill an adult. Smoke from burning oleander also produces poison symptoms.

Rain lily — a spring-blooming bulb that sprouts a slender stalk two to six inches long. Leaves are typically lilylike, slender, four to 10 inches long, and about a quarter-inch wide. Flowers are six-pointed petals, usually white or pink in color with six gold stamens in the center. The rain lily is cultivated by some homeowners, but is found in the wild in flatwoods, low grassy fields and swamps. The bulb is the poisonous portion of the plant. Poisoning symptoms: staggering walk, soft manure, sudden collapse.

Tung Oil Tree — found primarily in northern and western Florida, often along roadsides. This small tree has smooth bark and milky sap. Leaves are heart-shaped; flowers appear in the spring and are about one inch in diameter with five to seven petals in pink or white. The foliage, sap and fruit contain a toxic substance. Poisoning symptoms: animal has difficulty breathing, loses appetite, grows depressed, salivates, and its skin cracks.

Other plants, including some that are found in flower or vegetable gardens or are used as row crops, can poison horses if the animal consumes enough of the deadly meal. Among them are elephant's ear, larkspurs, allamanda, cotton, lupine, lima bean, buttercups, potato plants, chinaberry, elderberry and sorghum. Fortunately, most horses will not electively eat poisonous plants. Well fed horses who receive proper feed and hay will ignore the exotic goodies. The danger comes in winter, when pasture grasses are sparse and the owner has not increased rations to compensate. Horses then will forage on their own and may choose undesirable plants. It is also a well-known fact that horses who do not receive a properly balanced ration may eat all forms of undesirable material ranging from their own manure to toxic substances.

Following a few simple guidelines can reduce the chances of a horse becoming poisoned. They include:

- — Feed properly balanced rations in sufficient amounts.
- — Do not overgraze pastures.
- — Keep horses away from trash piles or waste dumps where many poisonous plants and seeds seem to thrive — or where such plants may have been discarded.
- — Keep horses off newly plowed fields where poisonous plant roots may be upturned and easily consumed.
- — Avoid turning horses out on or near dried watering holes, where plant roots may be unusually exposed.
- — Avoid stagnant water found in partially dry watering holes.
- — Keep horses away from around the house where poisonous ornamentals are planted or hung in decorative baskets.
- — Clear all pastures including the outer fence line where rubber-necking horses can stretch for "grass-grows-greener" goodies.

Persons who suspect they may have poisonous plants on their property should consult an agricultural extension agent to help make positive identifications.

An owner who fears a horse has eaten a poisonous plant should consult a veterinarian immediately. At-home treatment is risky and best left to a professional.

It is clear from noting the various symptoms associated with eating poisonous plants that poisoning isn't always easy to detect in horses. Then there are some afflicted animals who acquire a tolerance to harmful plants, or display only mild symptoms such as appetite loss or mild diarrhea. In addition, some severe poisoning cases may mimic other ailments such as encephalitis, colic or even rabies.

For an accurate diagnosis it usually takes an equine veterinarian's examination of the afflicted animal. But the owner certainly should suspect poisoning, or any other serious ailment, if a horse sways or staggers or stumbles around in circles. Other telltale signs of trouble are muscle spasms, slow or wildly racing breathing or pulse rates, biting or kicking at flanks or rolling on the ground. An owner may be certain the horse did not get into any poisonous plants but the animal, nevertheless, shows signs of ingesting some sort of toxic material. That is quite possible, for just as dangerously poisonous as plants are pesticides and insecticides.

Some words of precaution on that subject. Never mix pesticides or insecticides in any container that could later be mistaken as a feeder or a mixing implement. Moreover, it pays to be alert, especially in rural Florida, to know when neighboring farmers are spraying their fields. Of course, pesticide applicators are legally responsible for applying chemicals correctly but some cautious horse owners will, if seeing spraying underway from airplanes or other machinery, avoid trail riding in those areas for several days. Understandably, horses should not be turned out into adjoining pastures that may be contaminated by insecticides. Nor should cattle sprays be used on horses.

Another precaution is to dispose of, not reuse, empty insecticide containers. Some pesticide containers cannot be thrown into the trash because they contaminate landfills; Florida requires special disposal of such containers so checking labels is important. Just as toxic to a horse can be the evening's feed ration if it has grown moldy. Wet weather in regions where grains are grown can cause foodstuff to develop what veterinarians call mycotoxins. If a horse is mysteriously ill with no apparent reason, suspect poisoning in the feed and have grain tested per veterinary instruction.

A notorious poisoning case in Florida involved not a plant nor a pesticide per se but an insect found in a plant, specifically, in alfalfa hay. It happened in the fall of 1983. Imported alfalfa hay harbored dead blister beetles that killed two Ocala racehorses valued at $1.2 million. Both animals had eaten hay shipped from the midwest and southwest. Two Paso Fino horses in South Florida also died from eating the beetles.

The insect, though dead, contains an oily substance called

cantharidin. A substance harmful to all mammals, cantharidin in horses is especially potent. Reports say that as few as five or ten beetles ingested by a horse can cause death.

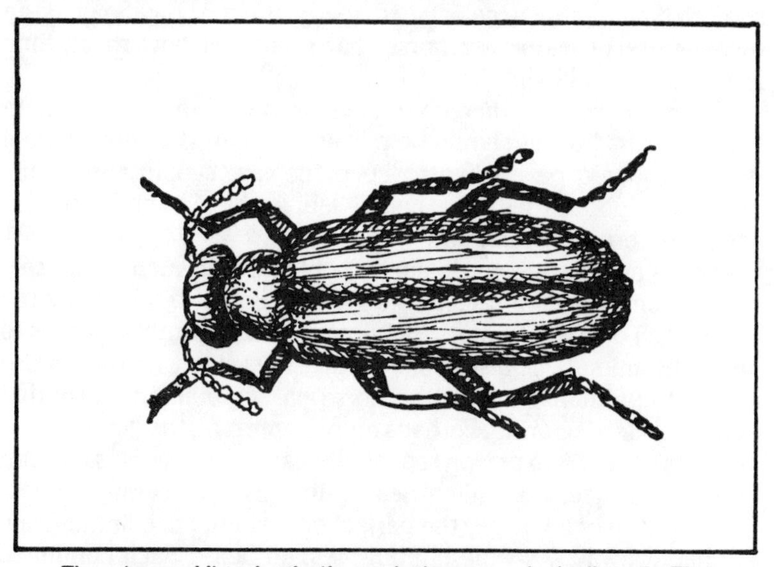

The poisonous blister beetle, if eaten by horses, can be fatally toxic. The beetles have been most often found in alfalfa hay. Before feeding any hay, carefully inspect it for these pests.

First signs of blister beetle poisoning are colic — in fact, early cases of such poisoning have been mistaken for a twisted gut. Horses pace and sweat. Within a couple hours (depending on the severity of poisoning), the horse's gums turn purple. In some instances the animal "plays" in its water bucket. This is because the mouth is irritated by cantharidin. Respiration and temperature increase, muscles start tremoring. Horses that survive the first 24 hours try vainly to urinate. Death finally occurs from severe irritation of the stomach lining, small intestine and bladder. Some owners who have endured the horrible death of horses by ingestion of blister beetles report their afflicted horses emitted a foul body odor. So far, alfalfa hay is the only edible found to contain the sinister dead blister beetle. Farm experts say the problem may arise because hay is now harvested by crimping or "conditioning" procedures, a process that prevents beetles from escaping from hay balers as they once did.

FLORIDA HORSE GROOMING

Grooming a horse involves three types of tidying: the all-around daily care, the occasional but more thorough cleanups, and the special beautifying necessary to ready for a show. So it is an owner's use of his or her horse that spells just how much time grooming a horse will demand.

Every horse, whether competed or not, requires minimum daily care. Its hooves should be picked clean and, if needed, hoof dressing should be applied to keep the feet supple and amply pumping blood through the horse's feet and legs. The horse also should be brushed each day.

There are reasons to do such chores beyond merely the self-satisfaction of having a clean horse. One is that by handling the horse each day, the owner gets "closer" to the animal — learns to know the animal's ticklish spots, the sensitive areas. An owner also is able to monitor whether the horse has suffered an injury that needs special doctoring. Perhaps more important, the horse gets to know its owner. As a person repeatedly handles the horse's feet and brushes its hide, the animal learns to trust this caring human. Grooming gives an owner the perfect opportunity to communicate — chatter away — to the big, four-legged beast. Such continued communication pays off later during riding and working together.

The first usual step in daily horse grooming is picking the feet clean of dirt and stones both before and after riding, or even if the horse has not been ridden that day. Next a curry comb is energetically rubbed over the horse's hide, down to the area above the knees. Currying is not done on the bony areas of the horse's body such as the lower legs or the face because the curry comb could rub the hair off these delicate places. Currying removes dead hair and excess dandruff, as well as dirty skin oils called scurf.

A stiff-bristled dandy brush is used to sweep away material that is loosened by the curry comb. A body brush can then be used, although some grooms like to use yet another brush, a soft dandy, before going to work with the body brush. The body brush has short, soft, close bristles that thoroughly clean the horse's coat. After using the body brush, the sheen of a healthy horse begins to blossom. A final wipe with a clean, soft rag or towel will bring out the natural oils in a horse's coat to make it glisten.

Florida's bot flies demand special attention. The fly lays its small yellowish eggs on the horse's coat. Warm water can urge the eggs to hatch and disappear. Occasionally, bots must be forced off. A block of wood with sandpaper attached is a quick way of gently rubbing bot eggs away.

Manes and tails do not have to be combed every day, but most owners do it anyway. Stalled horses seem always to get shavings or hay strands caught in their manes and tails. These can be pulled out gently. Care should always be used when combing the manes and tail. Most owners take pride in luxuriant growths of these equine hanks of hair, but manes and tails grow very slowly so every broken hair results in a stubby-looking horse. Some breeds are intended to have pulled manes or tails — but more about that later.

Basic grooming tools (clockwise): Mane pulling comb, regular mane comb, body brush, sponge, hoof pick, rubber curry comb, rubber wash mitt, sweat scraper. The large brush in the center is a "dandy" brush.

No owner wants to be kicked or nipped by a horse so the animal should always be confined during grooming. Unbelievably, some owners chase a horse around a paddock or a pasture during grooming, an intolerable scene. Three traditional methods of keeping a horse in one place are used. One is cross-tying. Cross ties are lengths of chain or rope that are attached to a stall, a grooming rack or an aisleway. The ends of the ties simply snap onto the

Tips and Quotes! A final rinse of one cup cider vinegar to one gallon of water makes the coat of a just-bathed horse glisten.

---

horse's halter. A well-trained horse usually cross-ties. Some horses truly resent being tied, and as soon as they are snapped they will lunge forward, rear up or pull backwards. Such a horse can hurt itself, break halters and seriously damage a barn. Some horses who do not cross-tie will tie by a slip-knot to a secure stall hook. While it is desirable to have the horse's head secured enough to keep it still, some horses are happy only if they have room to look around a bit. Non-tying horses can be taught to ground tie, that is, they think they are attached but the lead line is simply dropped on the ground and the horse stands quietly. A bitted horse should never be tied anywhere by its reins. Grooming should always be done with the same routine, the same procedures done in the same order each time. For example, when cleaning the horse's feet, always start with the same foot and work either clockwise or counter-clockwise. After a few times, the horse will have the next foot ready to be picked up. It takes new owners some time to figure out the most efficient regimen, but such extra attention to a routine makes a horse comfortable because it always knows what to expect.

As frequently as possible, a horse should be bathed all over. But be wary of household soaps. Combined with Florida sun, they will turn glossy horsehide into dull, dry wire. Manure stains can be gently washed off with a sponge. A damp, soapless sponge can be used on the horse's face. Separate sponges should be used for the horse's hindquarters. Sweat collects around the tail area and a gummy scum can accumulate that will invite horses to rub their tails. A daily wipe with bicarbonated soda on a damp sponge helps prevent the problem in this area.

When a bath is absolutely necessary, say, before a show, use a bucket of warm water with a soap product made for horses. These products usually include a lanolin or other conditioners that keep the horse's coat soft and shiny. Rinse the horse thoroughly to remove all soap.

Very special care must be used when cleaning male horses. Oils and other secretions can accumulate on the male sex organ and cause painful swelling. Occasionally, dirt can get inside the "sheath," an outer covering of the organ, and the horse will have difficulty urinating. Every horse is unique when it comes to being cooperative about this cleaning procedure. Some must be "twitched," meaning a restraining device is clamped around the horse's nose to make it hold still. A good safety tip: If the handler works from the left side, a friend can hold up the horse's left front leg. This offers some protection if the horse kicks up under its belly (a cow-kick). The horse cannot kick with its left rear foot or it will fall over; a kick with the right rear foot will likely miss the handler. Neophyte horse owners should consult with an experienced horse person, even their veterinarian, about when and how to clean a horse's sheath.

Any horse owner who intends to show — even if not right away — should get the horse accustomed to being clipped. In Florida, many horses never grow the large wooly-bear winter coats seen up north, so full body clipping may not be necessary unless the horse is going to be shown in rated shows from January to March. If a full clip is necessary, beginning horse owners should ask an experienced person for help the first time. Heavy duty clippers are used on the horse's body and neck; smaller trimming clippers, such as those used in barber shops, are perfect for a horse's legs, on the ears and around the face. Clipping a horse is a deceivingly big and messy job for the inexperienced groom. Itchy horse hair gets on and in everything. Dull-bladed clippers can gouge, even cut, the horse's skin. Blades also heat up and must be switched off, allowed to cool. Sharp blades will keep from pinching the horse's skin.

Horses, like people, need frequent breaks from such chores. A short walk and a rewarding bite of carrot or apple helps minimize equine fidgits. Being conservative is a good clipping rule. It's always easier to go back and touch up missed areas than to wince over every accidental but obvious bald spot. In fact, a poor clipping job can leave a horse looking as though it has been attacked by moths. Many Florida horse owners let nature take over the clipping chores. Spring's arrival prompts shedding of winter coats without having to resort to blades. This natural shedding can be assisted with use of a shedding blade, a curry comb or a soft rubber mitt that is covered in tiny, wart-like bumps. These mitts are flexible and gentle, so ticklish horses will stand still better for them than they do for the shedding blades. Grooms should always use the shedding blades with care because they can scrape the skin right off a horse, like using a potato peeler. And never use a shedding blade on the lower legs or face of an horse. Most owners will want to clip their horses' ears and keep a bridle path cut through the mane, behind the horse's ears. Such a path makes it easier to put halters and bridles on. Of course, different breeds require different widths of bridle paths. Arabian and Western stock breeds need a wider path cut than do hunters. Trainers can give advice for a particular showing style or for a particular breed. In some cases, the owner may "roach" the mane, that is, shave it off completely. Three-gaited Saddlebreds are shown with a shaved mane. And many Western owners like this style because it is less upkeep. Unfortunately, once a horse's mane has been roached, it may never grow out as nicely as it was originally. And, like most hair on the horse's body, the mane serves to keep flies off the horse's neck, so a missing mane affords less insect protection.

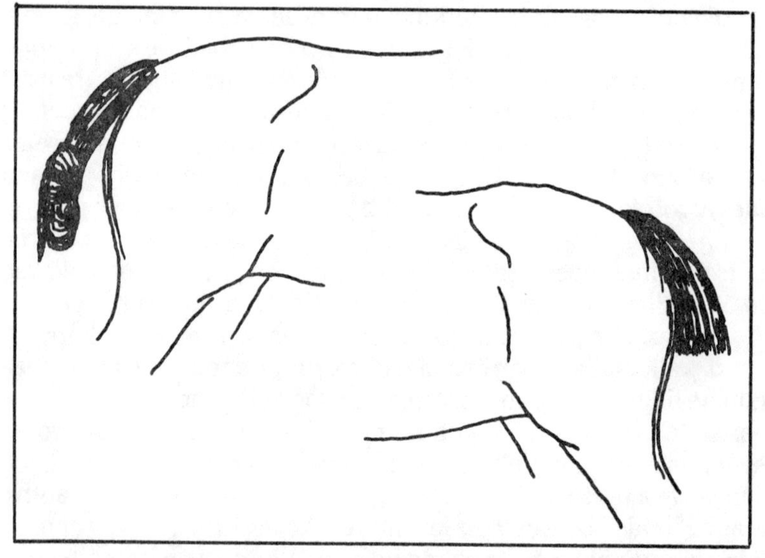

Tail styles: Mud knot (left) keeps tail clean in wet weather. Docked style (right), seen rarely and only on driving horses, prevents tail hairs from tangling with harness.

Corn starch, made into a paste, brightens the white markings on the horse's legs. And old-fashioned laundry blueing added sparingly to rinse water removes the "yellow" cast of a grey horse.

Similarly, the tail is a horse's chief fly whisk. Whereas it once was the style, especially for driving horses, to "dock" the tail — cut it off very short — that is less often seen today. Western show people do thin their horses' tails by pulling the longest hairs out. And Saddleseat or Walking Horse exhibitors often cut the tail tendons of their horses to keep their tails bannering high. These tails then are "set" in a supporting brace. Set tails need careful attention because the tail is a sensitive area in the horse. Novice Saddleseat or Walking Horse owners should consult with a knowledgeable trainer for advice on caring for the tail-set horse.

There are many ways to braid horses' tails and manes. How-to books that illustrate each braiding step with drawings or photographs are available on this subject. Basically, each style of riding has certain overall appearance requirements. Western horses are never shown with braided manes or tails. English Park-style versatility breeds such as Morgans and Arabians are shown unbraided. Three-gaited Saddlebreds are shown with a roached mane and with a set tail. Five-gaited Saddlebreds and Tennessee Walking Horses are shown with a long mane and a set tail. Pleasure classes in these latter breed shows often show without a set tail. English hunter-style and dressage horses are shown with a braided mane — if the braids make the horse look nice. A horse with a misshapen or scrawny neck looks nicer with the mane left unbraided. Manes do have to be thinned and shortened, though, to an even braidable length, about three or four inches long. This is done not with scissors but by removing the longest hairs with a pulling comb. Only a few hairs are pulled at a time and because horse hairs have fewer nerve endings than human hairs, the process is not painful, only uncomfortable. Braided or unbraided, the finished product should always lie on the right side of the horse's neck.

Tails of hunter horses are usually left about ankle-length and natural. A "banged" tail is one that has been trimmed off square at the bottom. This style has always been fashionable in Europe for racehorses and show horses and is becoming more evident in the United States, especially for dressage horses.

Chapter Five

Reining In The Sunshine

God forbid that I should go to any Heaven in which there are no horses.

— Robert Graham in a letter to Theodore Roosevelt

The object of owning a horse is to enjoy it. For some owners, that enjoyment may be merely watching a shiny horse contentedly nibbling at pasture. To others, the enjoyment can come from watching family members ride. But the vast majority of horse owners cannot imagine owning a horse and not ever climbing aboard it.

This majority generally begins riding by just playing around — not taking the riding business too seriously. In fact, most new horse owners are content just to avoid falling off the creature. Some of these backyard owners do not care to get fancy about riding. Others, however, get bitten by the lesson "bug."

It is safe to say that most Florida riders probably have begun riding in a Western-style saddle, with a horn that gives the shaky beginner something to hang onto. This type of saddle had its origins in Spain and came to the New World via Mexico.

Western-riding in Florida offers a variety of pursuits from simply working farm cattle to showing horses. And the state, despite its urban stereotype, teems with 4-H chapters that enjoy

Western riding activities. Add to them the scores of breed associations and local clubs that offer higher level "rated" shows and Western-riding enthusiasts could stay in the saddle indefinitely, or at least nearly every weekend the year round. The weekend association and club shows generally offer, among other competitions, pleasure classes that show off the easy-going manner of the classical stock horse. The rider is not judged. Riders' skills are keenly observed by a judge in what is called stock seat equitation classes. In Western trail classes, horses and riders must be unshakably calm throughout an obstacle course that includes opening a gate, passing through it and closing it; walking over a small bridge; backing through a short, L-shaped maze, and side-passing a log. Some trail classes involve obstacles such as a small fence that must be jumped. In the faster lane of Western showing are heart-stopping timed classes such as racing around barrels or pole bending — the equestrian version of a "slalom" through a course of flags. Interest in cutting horses also is increasing with new clubs forming regularly to exhibit Quarter Horse abilities to cut a steer out of herd. Show classes are scheduled even for people who do not ride. For them, the pride is in the horse's appearance as they proudly stand beside their horses in halter contests, where only the horse's conformation is assessed; or in grooming and conditioning classes, where the most spic-and-span horse gets the blue ribbon.

In a state where the largest number of horses are Quarter Horses, Western riding is the most popular style of riding. Coming in at a fast gallop on the popularity poll, however, is English riding. This is the "fancy" style of riding in which riders briefly rise from the saddle at the trot gait, although many Western fans grumble that rising looks silly — sissy, even. But cowboys who spend full time punching dogies in the Western wilds admit they also rise at the trot. The practice is done because it lifts the rider's weight off the horse at every other stride, briefly resting the animal's back and supporting diagonal legs. During a long trip, such as following cattle from one grazing spot to another, the cowboy will rise at the trot to keep his mount as fresh as possible. English riders rise the trot on their hunting horses for the same reason.

Other than the differences in riding garb, the principles behind riding, no matter what style, are the same: to get the horse to obey its signals and for both horse and rider to enjoy their riding activities.

English is divided into the two "sub-styles" of Hunt Seat and Park classes (also called Saddleseat). Hunt Seat is, as its name implies, the style of riding done in the foxhunting field. Hunt Seat is performed on the "flat" without leaping obstacles or going "over fences," and in classes that do involve jumping. In Hunter Hack classes (also called Under Saddle or Pleasure classes), the horse is judged on how easy going and relaxed it moves on the flat. In Equitation classes for Hunt seat, the rider is judged on how well he or she has mastered riding on the flat. A walk, trot, canter and hand gallop may be requested by the judge. Then there are classes for the Hunt Seat rider in which the horse's manner of jumping over fences is judged. In Over Fences classes marked as Equitation classes both the rider's performance in handling the horse and the horse's ability to jump obstacles in the proper manner are weighed.

Park horses are horses that commonly were ridden in public parks. When stopped, Park horses "park out," that is, they spread the distance between their front and back legs. This lowers the horse's back and makes the rider's mounting or dismounting easier. The usual Park style horses are American Saddlebreds, although Morgans and Arabians can be ridden Park style, too. All have high-stepping, animated gaits.

In addition to the basic English and Western styles of riding, horsemanship includes other activities such as endurance rides, in which style does not matter but how quickly the course is finished does count. Such courses may vary from five miles to 100 miles long. Too many tragedies occur on endurance rides in Florida because riders enter horses that are not conditioned for such extended trips. Even on the coolest of days, most backyard or pleasure horses are not fit enough, unless regularly exercised, for such a long grind. More than one horse has dropped dead on such rides, its heart or organs unable to perform under the stress. In Florida, endurance rides are particularly challenging if held in the summer months. The heat and humidity take the punch, if not the life, out of otherwise healthy horses during an endurance trek. This is not to say a horse and rider in Florida should not enjoy endurance rides. But it cannot be emphasized too strongly that a rider should make sure his or her horse is up to such a test, particularly during the draining heat of summer.

Horses also must be fit to participate in Combined Training, also

"In riding there's a physical and a cerebral way to do anything. I believe in the cerebral way. For instance, I've never understood why — when people have more brains than horses do, and horses have more strength than people do — why people don't just outsmart a horse. Instead, people try to outpull a horse — and the people lose every time." Helen Crabtree, Saddleseat trainer

---

called Three-Day Eventing. Eventing requires a horse to endure a dressage test calling for intricate maneuvering on the first day, followed the next day by a miles-long cross country run over hedges and fences, and ending the competition on the third day with a round of "stadium" jumping — hurdling a course of five-foot-tall (or slightly bigger) fences. Eventing is perhaps the equestrian world's most challenging test for a horse's stamina and ability. It should be entered only by the fittest of horses.

Dressage — performing a routine of turns and circles at every gait according to a prescribed (memorized) test — is not merely an eventing contest. Dressage clubs abound in Florida and their members concentrate only on dressage competition, leaving cross country runs and jumping to other enthusiasts. Dressage tests, at many levels, are held on the flat in a rectangular arena about half the size of a football field.

Open jumping is another name for stadium jumping in which the horse and rider must negotiate a course of tall fences. Such jumper classes are not restricted to a certain breed of horse, but riders use English jumping saddles and wear English Hunt Seat style attire. The form a horse uses to get itself over the fences does not matter, only whether the horse knocks down any portion of the obstacle, which earns penalties called "faults." The horse and rider with fewest faults wins.

INSTRUCTION

Instruction in either Western or English riding styles is readily available anywhere in Florida. And for all age groups and types of people. Classes at some stables are available for the young pony crowd, instruction can be found around larger cities for the handicapped and/or mentally retarded, and clinics directed by professional riders of national reputation are scheduled throughout

The American Horse Council in Washington, D. C., says horseowners spend a whopping $13 billion dollars to feed, shoe, transport and house their equines each year.

---

the state. It is not necessary to own a horse to participate in classes. In fact, it is a good idea to take lessons before purchasing a horse; a person who becomes basically proficient in riding knows better what type and size horse is best for him or her.

Most large stables provide riding lessons. In addition, individual teachers at smaller stables often advertise that they give riding lessons, either at their stable or at the student's farm or backyard. For those who look to riding lessons as part of their formal education, many private and public schools offer equestrian courses. The University of Florida, for example, gives instruction and has its own equestrian team that competes against other university teams around the country.

Costs of lessons vary considerably. Beginners receiving individual instruction can expect to pay a minimum of $10 a hour. This fee can go as high as $30 an hour depending on the teacher and quality of instruction. Clinics are excellent for all students because the instructors are often some of the best in their field and because instruction is concentrated, that is, the clinic extends from one to three days and generally is limited to a specific activity such as dressage, eventing, jumping or a facet of Western riding. Clinic costs can vary tremendously, $50 to several hundred dollars. Non-riders also are welcome at most clinics and pay a much smaller fee to "audit."

The beginning rider with a small pocketbook should consider enrolling in group lessons. All that means is a teacher instructs, say, two to five students simultaneously. Each student pays a reduced hourly fee, sometimes half of what it would cost for a private hourly lesson.

Because Florida is such a popular equestrian state, the number and quality of instructors available is vast. And near large cities — Tampa, Miami, Jacksonville, West Palm Beach, Tallahassee — more and more equestrian centers are being built, each offering not only instruction but riding facilities and weekend competition. Such facilites often offer video films of lessons, something that can be particularly helpful to a beginner. On a horse it is impossible to notice one's own form, what one is doing properly or improperly. Video shows it all. It is a valuable teaching aid.

A word of advice: Riding is little different from tennis, golf or any other sport that requires daily practice. A student who owns a horse will derive greatest benefit from lessons if he or she practices between classes.

SHOWS

There certainly is no dearth of horse shows in Florida. State magazines and newspapers devoted exclusively to riding fill pages in each issue with a "Calendar of Events." These events are horse shows open to nearly anyone who has a horse. Still more pages are devoted to listing of winners and top placers in previous recent shows, a list that often includes thousands of names.

Nearly as impressive is the variety of shows being held, everything from Western to breed shows for Paso Finos, Buckskins, Appaloosas, Arabians, to the gamut of English Hunt Seat and Park-type contests. Open shows are those offering the widest variety of competition. Some two-day shows offer everything from Park classes to Tennessee Walking Horse classes to Paso Fino classes to Jumper classes. In addition to this wide variety, open shows also lend experience to a range of riding abilities. They usually include such classes for beginning horses as Novice, Baby Green or Green classes. Junior classes are available for younger riders (18 years or under, usually).

Anyone dedicated to showing should join the American Horse Shows Association and get the organization's rule book. It details different types of shows, their regulations and their rating system of A, B or C. "A" shows are the highest caliber with the most formidable competition and, usually, the top judges.

Cost of entering classes can vary widely, too, from $3 at small, unrated local shows, to hundreds of dollars in "A" rated championship classes.

Florida is fortunate to have the United States Equestrian Team members visit the state for about two months of competition during the winter. The Florida Winter Circuit showcases the nation's top riders in the disciplines of stadium jumping and dressage competitions. Local Hunt Seat riders are offered a variety of flat, equitation, and over fences classes at these annual Circuit show sites in West Palm Beach and Tampa.

EQUIPMENT

Horses in Florida, like elsewhere, need certain basic equipment. First on the list is a stall halter. It may be made of nylon or leather and usually has a matching lead shank. Stall halters should be removed when the horse is in its stall or out at pasture because an animal can become snagged on fencing or stall furnishings. Long-lasting nylon web is the usual choice of most horse owners and costs from $8 up. These halters come with and without suede linings on the inside of the nylon. Both types get dirty, and the lined variety has a tendency, when permeated with horse sweat, to rub the hair off a horse's face. For $15 and up, leather halters can be purchased. Still another style often seen around Florida barns is the rope halter, an unusually sturdy type that costs about the same as nylon web halters.

Horses shown at halter must have a show halter. These vary in style and quality, of course, depending on the type of horse and the style of show. Thoroughbred hunters generally appear in a plain leather halter. Stock horses often are shown in leather halters usually decorated with some silver trim. Most often shown in a lightweight halter are versatility breeds such as Arabians and Morgans. In the case of Tennessee Walking Horses and Saddlebreds, they usually are shown in their bridles after their saddles have been removed during a class.

Whatever the style of horse or exhibition, there are, nevertheless, several halter basics. For one, the halter should fit properly, never appearing bulky or too small. It should be of a good leather and should complement the horse's appearance. In addition, it is best that the halter's lead shank match the halter, and that the halter have quality fittings, those metal parts that hold a halter together. Stainless steel and brass are usual.

Good halters generally cost $50 and up. One special halter, designed for Arabian horses, is made of gold and silver threads and costs more than $400. Silver-laden Western show halters likewise can cost $500 and up.

Next most important equipment for any horse owner to purchase is the bridle and saddle.

As with a halter, the bridle needs to fit properly. The bit should be chosen with care; it must coincide with the type of riding to be done. Bits come in sizes, too, with a size five fitting the average full-size horse. Smaller breeds, such as Arabian horses and ponies will need "cob," "Arab," or "pony" size — usually four or four and a half.

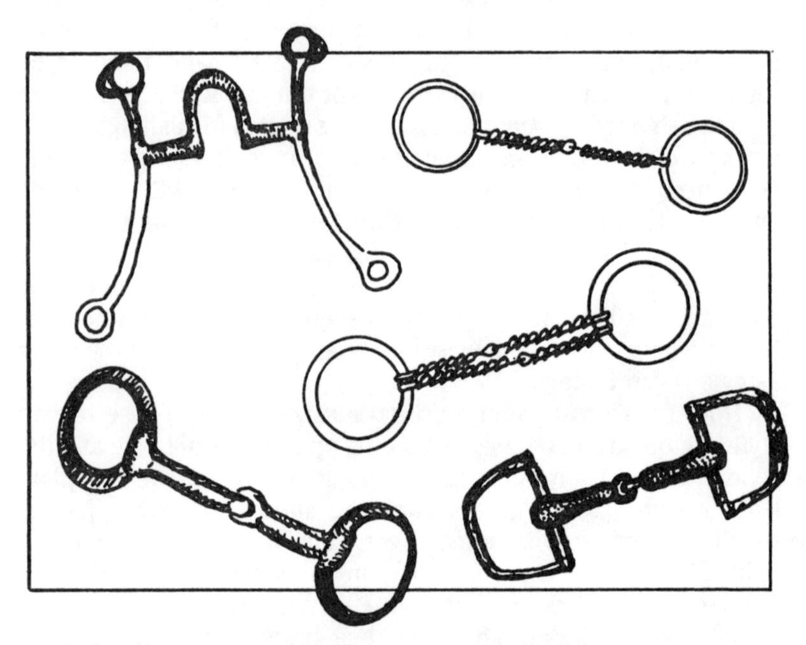

A wide variety of bits are availble to horse owners. It may take considerable trial and error to find the one most effective and best for your use. Pictured (clockwise from upper right): Single twisted wire snaffle, a severe bit; double twisted wire snaffle (also severe); rubber D-ring snaffle (mild); eggbutt snaffle (mild); high-ported correction bit (severe).

When selecting a bit, horse owners may want to consult with their instructor or another experienced horse person. Many errors are made by novice owners when buying bits. The premier mistake is buying too severe a bit whereas the goal should be to obtain as gentle a bit as possible. Rule of thumb generally is: The thicker the bit, the kinder it is. Twisted, sharp-edged or pencil-thin bits all are considered harsh. And the bicycle chains that are often advertised as "controlling" bits should be left to the pros; in the hands of an unbalanced novice, these bits become torture devices.

Tips and Quotes! The dishwasher makes easy and thorough cleaning of bits, stirrup irons and their rubber pads — placed preferably on the washer's upper rack.

Bridles do not always come with reins and that has an advantage — separately purchased reins permit the rider to choose the length and style that is most desirable. Eventing and hunter riders, for example, often select reins with rubber areas that make slipping less likely. No piece of equestrian equipment is as expensive as a saddle. Indeed, it is a breeze for the saddle shopper to spend four-figure amounts of money. Expense is usually linked to the intended use of the saddle. For example, a beginning rider, either English or Western, need not invest in show-quality equipment until he or she is ready to take the competitive plunge. Plain Western saddles of reasonably good quality are available from $100 up. Most beginning riders start with a Western saddle because it offers more places to hold onto — although no saddle is going to guarantee that a person will stay on a horse. In fact, many falls occur because the saddle comes off the horse, often because it is not properly affixed.

An average Western saddle weighs 35 to 45 pounds, generally more than an English-style saddle, but ladies' and childrens' Western saddles are available in lighter weights. The sky's the limit on cost, a new, highly decorated or custom-made Western saddle running more than $5,000 if it includes such amenities as a suede seat, sterling silver stirrups and handtooled designs.

Because English saddles do not have tooling or silver and contain less leather than Western saddles, they generally cost less. Top English saddles go for about $1,000.

Many beginning English riders, particularly Hunt Seat enthusiasts, buy Argentine-made saddles or saddles made in India or Japan. These "generic" or "all-purpose" pieces of equipment are advertised in retail catalogs and seem to be quite a bargain, rarely priced over $150. But cheap saddles may be false economy. If made of inexpensive grades of leather, they often do not wear well, are chronically dry and crack easily. These saddles also never become comfortably broken in and may have been put together with screws and other pieces of hardware that cut into the rider's leg or

"Recently it has become popular to use running martingales, draw reins or even tie-downs attached to curb bits. These are disasters waiting for a chance to happen. Horses in this kind of device become mad, scared or frustrated and throw a fit, usually resulting in a serious wreck. This is especially dangerous for young or inexperienced riders." Dr. R. D. Scoggins, veterinarian, University of Illinois

---

seat. Moreover, the inexpensive saddles are generally assembled around a "tree" that is bare wood with little padding. As a result, the "tree" cracks easily, rendering the saddle dangerously useless, not to mention pinching saddle sores on the horse's back. Some riding instructors complain, too, that these saddles are still constructed on an old, out-of-date pattern that places the rider's legs too far forward.

Under every good saddle fits a good pad. Western owners ride with a "blanket" under their saddles. It may be plain or have a colorful Indian pattern woven into it. Many Western riders find that placing a smaller blanket under their larger one keeps the larger one cleaner. English riders can select from a variety of imitation fleece-type pads. Although colors are available, it is always best to stick with white or the off-white pads. Pads, whether English or Western, should be kept clean and should neatly fit under the saddle.

But back to saddles. The best, regardless of style, are made of good grained leather that has been properly tanned and stoutly stitched onto heavy-duty webbing around a first-rate wooden or leather "tree." Some saddles even feature water-proof leather.

Newcomers to the equestrian world may not know that most English saddles come without stirrups. These fittings, as they are called, are extra. The stirrup leather and the "irons" are purchased in accordance to the rider's leg and foot size. Good quality leathers are always desirable because they tend to stretch less under the rider's weight. Irons are made in either nickel metal or stainless steel. The former is a greenish colored metal that always looks dirty. Stainless steel irons are prettier, easier to keep clean, slightly heavier — a plus when recovering the devils if they slide off the foot while riding.

No beginner should endure the misery of breaking in a new

Soften water-soaked leather boots or tack with a light application of kerosene.

---

saddle — a process of using, cleaning and oiling a saddle so that it begins conforming to the rider's seat and legs. Used saddles are generally supple enough to quickly conform to the rider's anatomy, are less expensive and are an all-around better buy for beginners. Moreover, saddles maintain their value and usually can be resold for their original purchase price — sometimes more.

When purchasing used tack, including the saddle, always check the stitching carefully. Straps that attach to girths or cinches should be especially eyed closely for cracks and splits; nobody needs an accident caused by a broken billet or cinch strap. Girths are the means of attaching an English saddle to the horse, and most often made of leather, nylon cord, canvas or mohair; cinches are the Western equivalent, usually fashioned from cord or mohair. Imitation leather girths sometimes are made of vinyl, a material that gets quite slick once it has been on the horse a few minutes. The advantage to nylon and mohair, as well as cord, girths and cinches is that they are washable, lightweight and cooler for the horse to wear.

No matter what expense has been incurred to purchase a halter, bridle, saddle and other leather tack, that money is down the proverbial drain if the equipment is not kept clean and dry. Florida's heat and humidity quickly reduces leather to shriveled and dry crispiness, or coats it green and fuzzy with mildew.

Needed is a good saddle soap. Many are available — in sprays, pastes and bars, with or without glycerine. All work well (most horse owners find their favorites and are eager to recommend them to novices) as long as they are used with a minimum of water. If the saddle soap sponge is lathered, there is too much water. A good oil should be applied after soaping the leather. Again, all horse folks have favorites, ranging from chicken fat to olive oil to manufactured oils such as Neatsfoot. In Florida, if saddles and bridles are not oiled after cleaning, they will harden and mildew. Likewise, if too much water is used during cleaning, it will invite

Some Florida horse owners opt to blanket their horses on chilly nights. After the cold weather is over, dry clean the blanket and store it away in a plastic bag to keep unwanted insects or rodents from nesting in it.

---

mildew. But be sparing — if oils and conditioners are not used with a light hand they will rot stitching and also attract insects.

Horse sweat, as well as heat and humidity, takes a toll on leather goods. The sweat and dirt combine to form black pimples that some oldtimers called "jockeys." These sweat bumps harden until stitching or decoration is obscured. And once soap is added to these areas, the jockeys turn into mini-mudpies that are gummy and disgusting to remove. Eventually, the old dirt will rot stitching and get anything they touch filthy. This whole problem can be avoided easily if the user takes time to faithfully wipe off the saddle with a quick saddle soaping after each ride. Of course, a clean saddle pad helps keep the saddle clean. Florida's cockroaches also can greatly damage leather. It only takes one encounter with roaches to reduce a lovely saddle, bridle or halter to a maimed and scarred shadow of its former self. Roaches will eat off the top surfaces of the leather and leave the soft middle layers exposed. The exposed area is weakened, and can never again be cleaned to a gleaming shine. Rats and mice do similar damage.

To properly clean the bridle, it must be disassembled. For beginners, all those straps and buckles can be a nightmare. A tip: Either make a quick diagram of the bridle, or number each piece with masking tape before and after cleaning.

Tack should always be stored in clean and dry areas, preferably away from feed that attracts leather-eating vermin. If equipment cannot be stored anywhere that bugs or rodents won't attack, then it should be stored in a bag. Tack stores and mail-order firms offer storage bags for saddles. Storing tack in a car trunk may be convenient but is not a good idea. Some trunks leak when it rains, ruining leather. Trunks also get very warm and high heat severely dries out leather.

There seems to be no end to horse equipment, but many of the items need not be purchased right away unless a student is instructed to do so by a trainer or instructor. One item, however, that every horse owner should have from the start is a longe line. These long 15- to 30-foot lines are intended to keep a horse going around in a large exercise circle. They cost between $6 and $20. Longe lines are also useful for teaching a horse to respond to verbal signals, important in early training. But young horses can incur leg problems if longed too long or too strenously; keep longeing sessions to no more than 15 minutes or so during the first few months.

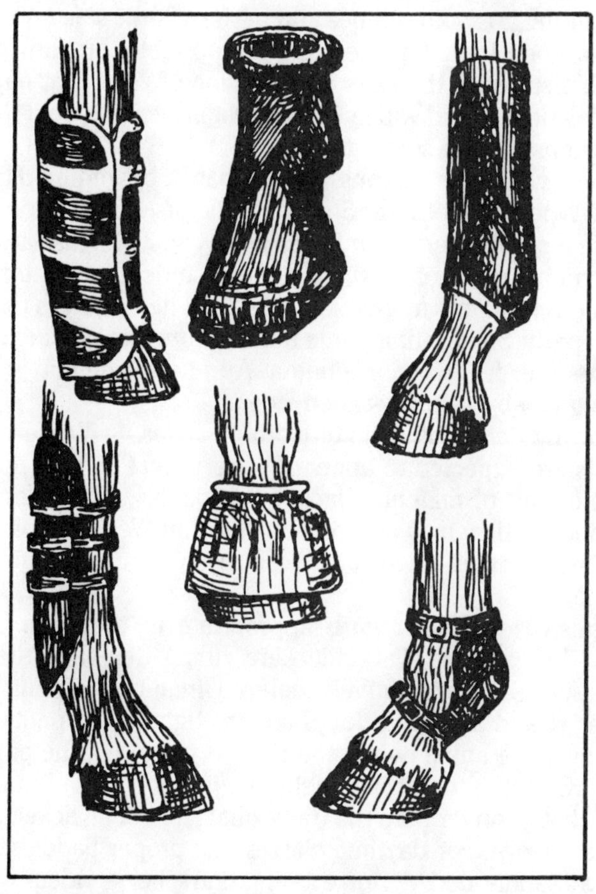

Not just riders wear boots. Equine boots serve a multitude of purposes. Pictured (clockwise from upper left): Washable shipping boot; rubber poultice or soaking boot; supporting galloping or splint boot; protective skid boot; supporting open front jumper boot. The rubber bell boot (center) prevents the horse from nicking itself with another foot.

CLOTHING

Beginning Florida riders should not wear fancy duds, just functional ones. In the state's warm climate, it is tempting to fun around on the trails wearing shorts or even a swim suit. Such a practice is not too smart, however, for even a laid-back backyard horse can accidentally step on its owner's sandal-clad foot — or scrape too closely to a gnarly old oak tree.

Anyone riding a horse should wear long pants, preferably snugly fitting, to avoid scrapes and blisters. Some safe type of shoe is also a good idea. Sneakers are comfortable, but can go right through a stirrup if the rider loses balance. When that happens a rider may be dragged with a foot caught in a stirrup until the horse finally stops to grab some grass.

Riders who take lessons will probably be required to wear certain types of clothing. Some instructors demand only good fitting jeans, boots and a comfortable shirt while English instructors may require breeches, and for jumper students, protective hard hats. Such clothing is more than stylish, it is necessary to ride safely and properly. Snug-fitting pants not only prevent chafing but they help the rider feel the horse better. And horses definitely feel the signals given by a sturdy-shoed rider.

Showing demands definite clothing styles. In Western classes entrants are expected to appear Western, that is, they may wear a frontier suit of matching jacket and pants with a Western hat and boots, or they may opt for jeans with an Western riding jacket or no jacket at all, merely a Western style shirt. Chaps usually are optional.

English riders wear garb appropriate to the style they are riding. For example, Park Pleasure riding demands the classic "saddlesuit," a conservatively colored man-tailored suit coat, a plain shirt and jodphurs. Jodphurs are tight-fitting pants that go down over the ankle (when the rider is mounted, the pants cuffs should reach the top of the heel). The jacket is form-fitting with a "skirt" that is longer than the traditional Hunt Seat jacket. Jodphur boots are worn. For daytime classes, the proper Saddleseat show hat is a "saddle derby"; for evening, Park horse riders wear tuxedoes and either derbys or silk top hats.

Hunt Seat riders wear breeches instead of jodphurs. Breeches are shorter and are worn with knee-high boots. A plain shirt is worn under a conservatively colored and form-fitting hunt coat.

The current key to dressing for show is to go conservative. Flashy colors, and gaudy boots or hats are not in style with most judges. Browns, dark hunter green, navy blue, and black are always correct color selections for hunt coats or for the Saddleseat suits (coats and jodphurs match). Breeches should be buff (beige), canary (pale yellow), rust or black. Hats, whether Saddleseat derbies or velvet hunt caps, should be in either brown or black. Although backyard "unrated" shows allow more flexibility in colors that riders may wear, economy-minded exhibitors who intend to proceed into rated shows are dollars ahead to initially invest in the correct attire.

Materials used for riding clothes have come a long way. Polyester has made for stretchier, more comfortable fitting riding clothes. For that reason, many newcomers to Florida put their woolens into storage and rush out to purchase "cooler" outfits made of "easy care" synthetic material. Of course, wool is too warm for Florida's summer competition, but the state's winter months are brisk enough to warrant wools. And many riders are surprised to find that polyester is not really as cool as cotton, simply because cotton is absorbent and "breathes." Cotton/polyester blends are a good compromise.

Show clothing comes in a variety suited to particular types of riding styles. Pictured from left is the matching Western suit; the Western shirt and pants worn with chaps; the Saddleseat suit worn in Park-type English shows; a dressage outfit with long-tailed "shadbelly" coat; the Hunt Seat garb worn when riding hunters or jumpers.

Breeches are commonly made in synthetics and the "fancy" labels of breeches can cost $100 per pair and up. These are available through many catalogs, but it is always a good idea to go to a tack shop and try on a variety of breeches. Some synthetics are durable but very thin and form-fitting, like leotards. White breeches are especially "see-through," and some riders may feel "over-exposed" in such revealing pants. Female riders should always have their hair pulled back in a knot or pulled up under a hat. The rule for jewelry such as necklaces and bracelets is to save it for later. Dangling earrings are likewise frowned upon. A pin worn on a choker collar or on a stock tie is the only jewelry most judges care to see. That is because such traditional pins, and stock ties, have a history based on utility: In the hunt field, pins and stocks could double as an emergency bandage. If riders have any doubts about proper attire, they should check with tack shops, trainers or with their American Horse Shows Association rule book.

DRIVING YOUR FLORIDA HORSE

Florida horse owners are discovering that an exceptionally relaxing way to enjoy their equines is to drive them. Clubs around the state have formed to hold driving competitions and social get-togethers.

Driving a horse behind a cart or buggy is not only used as an equestrian activity, it also is a handy way to train a two-year-old that is not yet ready for under saddle work. In fact, young horses that have been driven are often better balanced and better behaved than their undriven counterparts. Moreover, even a pony can pull an adult or children in a lightweight cart. Some inexperienced horse folk have had success training their own horses to drive, but it is best done under the watchful eye of an experienced hand. The logical steps in beginning are for the horse owner to learn about properly harnessing a horse and driving — before launching any horse training missions. As with any equestrian endeavor, certain basics must be observed to keep from injuring horse or human. Besides, carts and buggies can crash and repairs may take dollars and months to accomplish.

Unlike riding, the driving horse must get used to pulling weight, to the feeling of wooden or fiberglass shafts that inhibit its movement, to the ever-present vehicle close on its heels.

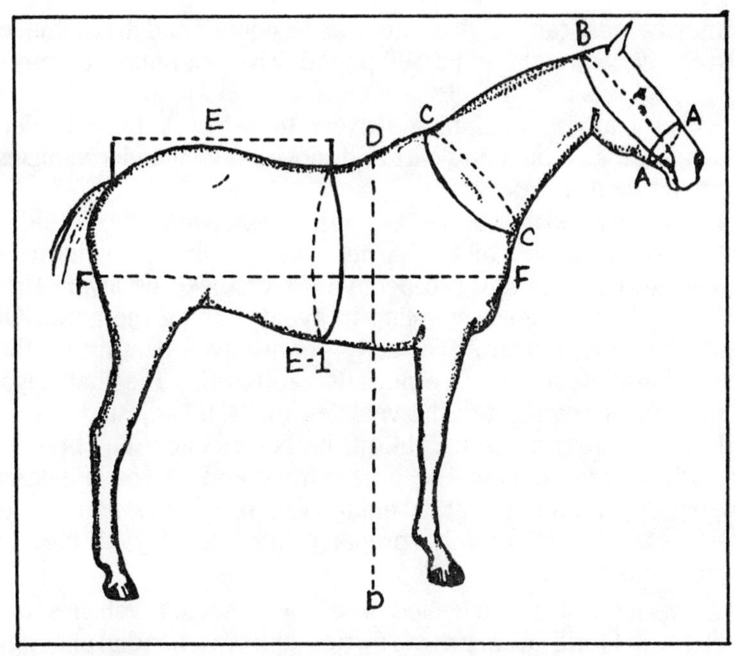

Measuring for a driving harness should include all of the indicated points. The point "F" to "F" also is the line measured to fit a horse blanket.

Horses are strong creatures. History reveals that six artillery horses pulling a half-ton gun could keep pace with any cavalry unit. However, as with everything else to do with horses, the British have rules on this matter of pulling. Basically, the rules say that a horse should be capable of pulling its own weight, literally — hence the old saying. But this is conditional. For example, the horse or pony must be in good condition and properly harnessed. The carriage or cart must be in good shape and freely-running. Pulling power also depends upon the breed of horse and its conformation. Thus, a 10-hand pony would be asked to pull no more than about 330 pounds; a 16.2-hand Cleveland Bay (a hefty breed) could manage about 1,300 pounds. Of course, other factors such as the weather, footing and terrain also make a difference in a horse's ability to manage a vehicle and its passenger(s). When figuring the weight a particular horse will be asked to pull, consider more than the weight of passengers. There is the cart itself, even the the harness.

A fancy harness can weigh as much as 16 pounds and it, combined with the cart's weight and a 300-pound driver can make a pony's job a tough one.

Experts advise beginning drivers to select a two-wheeled vehicle such as a breaking cart and move up to showier vehicles and four-wheelers later.

Recommended harnesses are lightweight with "furnishings," that is, the metal parts on the harness, made of brass. The harness should not be fussy and prospective drivers need be aware that such new leather needs suppling to avoid rubbing the horse. Of course, all pieces must fit correctly — with even bearing on the horse's body. Most experts especially caution that the shaft tugs, which carry the weight of the vehicle's shafts, be adjusted evenly and not exert too much weight on the horse's back. The breeching, which goes behind the horse's rear end, is the breaking power; measurement of this equipment piece, say experts, is often neglected. It must be properly fitted to play its part in stopping the vehicle.

A variety of two-wheeled and four-wheeled vehicles are available in Florida from private owners or from a handful of buggy makers. Used, even antique vehicles, may be purchased for $500 and up. These are usually not restored, however, and must be checked by a knowledgable driver. Many driving enthusiasts, however, order equipment from national manufacturers. A popular source for harnesses is Amish harness makers. A favorite show for Florida driving enthusiasts, and spectators, is the annual Winter Carriage Driving Festival held in February-March in Palm Beach and in Tampa. Competition includes classes for single horses, pairs, and fours-in-hand. Horse breeds that may be seen include everything from ponies to draft horses. Horses and vehicles vie in obstacle courses, marathons and reinsmanship classes. Culmination of the Carriage Festival is Tampa's Concours d'Elegance class in which the horses and vehicles are judged for elegance. In this class the judge scrutinizes the vehicle, harness, horses, driver and passengers. In fact, the vehicle also may be inspected as to how well it has been restored.

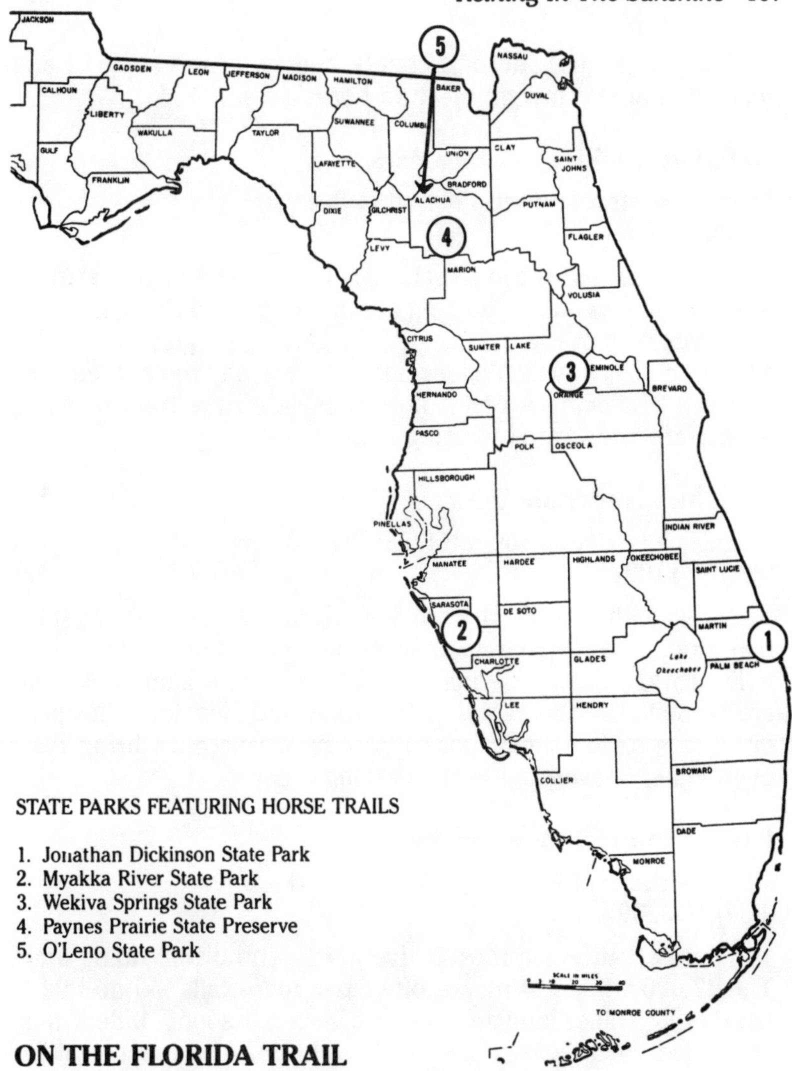

ON THE FLORIDA TRAIL

Florida's State Park system celebrated its 50th anniversary in 1985 and it is one of the finest such systems in the nation. In fact, most Floridians would attest that the "real" Florida is no longer found on the condominium-lined beaches, but in the state's 90 state parks, recreation areas, preserves and other recreation areas that are being jealously protected. Naturally, horses are increasingly considered in the future of such areas. In fact, five state parks as well as dozens of state and county operated forests, parks and recreation areas currently feature horse trails.

Jonathan Dickinson State Park

13 miles south of Stuart, Fla., off U. S. 1
(905) 546-7199

This park offers the longest trails, 13 miles of them, that wander through pine woods to Kitching Creek. Sharp-eyed riders may spy rare birds such as the bald eagle. The park offers guided rides for a fee. Horses are available for rides, or riders may trailer their own horses in. Drinking water is not available on the trail, so horse people are forewarned to bring it along.

Myakka River State Park

17 miles east of Sarasota off State Road 72
(813) 924-1027

Myakka River State Park provides 12 miles of scenic riding trails that loop in three separate areas of two miles, four miles and six miles. Parklands include marshes, flatwoods and hammocks that are home to bobcat, eagles, deer, turkey and alligators. This park does not provide riding horses, so equestrians must bring their own. Water is available in the parking area.

Paynes Prairie State Preserve

10 miles south of Gainesville off U. S. 441
(904) 466-3397

Paynes Prairie is a most active north-central trail riding area. The 17,000 wildlife preserve offers two loop trails — one that is nearly three miles long and one that is six miles long. Riders must bring their own horses, but the Prairie does have a corral for unloading and saddling horses. Reservations are recommended.

Wekiwa Springs State Park

Three miles northeast of Apopka off State Road 436
(305) 889-3140

This park boasts that it has changed little since Florida's Indian tribes roamed the territory. Horse trails wind through eight and one-half miles of flat woods that are home to a wide variety of animal life. Horses must be trailered into an area near Sand Lake. No water is available on the trail, so riders are advised to bring some along.

O'Leno State Park

20 miles south of Lake City off U. S. 41
(904) 454-1853

Riders aboard their own horses can enjoy eleven miles of trails across open fields and hammocks in the River Rise State Preserve area of O'Leno State Park. Two looping trails, one four and one-half miles long and another that is six and one-half miles long, provide the Park with a 20-horse capacity. A barn that once was the old McCleod homestead is used as shelter during a sudden rainstorm. Park officials advise riders to bring their own water.

The state parks above are open daily from 8 a.m. until sunset. Nominal entry fees are charged at these parks. Firearms, dogs, alcohol, fires and trapping devices are prohibited. Moreover, riders and horses are confined to the marked trails and may not ride out of these areas. In addition to the state parks mentioned above, locally operated trails are spread throughout the state. Persons who want more information about any available trails in their immediate areas are urged to contact local riding clubs or the Affiliated Horse Organizations of Florida (see Directory).

Rural and park bridle paths are safer than areas that are dangerously near traffic. Still, riders must be aware that Florida harbors a few rarely seen wildlife dangers. Of these, the chief concern is snakes. Fortunately, horses rarely die from poisonous snake bites. The horse's body weight — in proportion to the snake's venom — provides a safety margin. The exception, however, is if the horse is bitten on the muzzle, on the head or on the neck. These areas are closer to the brain. Nevertheless, horses attacked by poisonous snakes should be given medical attention as quickly as possible because snake bites can cause tissue damage in the affected area. If stricken on the leg, the horse often becomes lame.

Following are the principal poisonous snakes that could be encountered on the Florida trail.

Eastern Diamondback Rattlesnake

This largest of Florida snakes is a pit viper whose claim to fame is its rattling — a noise that sounds surprisingly loud and resembles a locust's buzz. Rattlesnakes generally are about three feet long, but can reach seven feet or slightly longer. Generally, rattlers are timid and will avoid people and horses. However, rattlers, while sunning themselves, sometimes do not hear approaching footsteps. When startled awake, these snakes may back off or may stand their ground. They can strike approximately half the distance of their length, but that is hard to determine when the snake is coiled. Horsemen should quietly and slowly back their horses away from a rattlesnake — and avoid panicking. Florida rattlers may occasionally seek refuge in rabbit or gopher turtle holes, so riders should avoid these. Never allow a horse to sniff down inside such hollows.

Canebrake Rattlesnake

This member of the rattler family is frequently found in Florida's flatwoods, river bottoms and abandoned fields. In hot weather, it sometimes retreats to swampy areas. Compared to the Eastern rattler, the canebrake is smaller, skinnier and has a pinkish or greyish coloration. It also has more vivid coloring down its back — an orange or rusty-red stripe. It is rarely found in central or south Florida and it is usually less than five feet long.

Pygmy Rattlesnake

This tiny, gray rattler produces a loud buzzing noise if disturbed. It strikes quickly, but rarely fatally, especially to horses. Pygmy rattlers often are found around marshes, ponds and lakes, but also live in palmetto flatwoods or pinewoods.

Cottonmouth Moccasin

The Florida Cottonmouth is generally only two to four feet long. It is sometimes called the "water moccasin" because its favorite hideaway is around swamps, lakes and rivers. Sightings are usually of snakes that are sunning on logs, stones or at the water's edge. These snakes, like others, usually opt to retreat slowly from horses and riders, but some will stand their ground. The cottonmouth opens its mouth wide and exposes a cotton-white interior, which is the basis of its name. Several types of non-poisonous water snakes resemble cottonomouths but a human can tell the difference by whether the snake vibrates its tail. Cottonmouths do vibrate; non-poisonous water snakes do not. The cottonmouth is found in all areas of Florida except the Keys.

Copperhead

Just a few northern Florida counties, primarily in the Panhandle, are home to copperhead snakes. The snake's name is somewhat of a misnomer in Florida because the species is paler, less coppery-colored than its northern cousin. It is usually found in lowlands near swamps and cypress stands.

Coral Snake

This is the prettiest of Florida's poisonous snakes with wide bands of red and black divided by thin lines of yellow. The coral snake has the most virulent venom of any Florida snake, but is also shy and retiring, preferring to hide under leaves and in old logs. It favors the dense "jungle" hammock areas and generally moves about in the daytime. Coral snakes are rarely more than two feet long. Most bites from these snakes occur when they are picked up and handled. The harmless scarlet kingsnake is often mistaken for the coral snake. Caution would dictate both species should be left alone; the snake that is marked with yellow coloring exactly next to red is the poisonous one. A little saying, "red and yellow can kill a fellow," helps the trail rider to make the quickest identification.

Florida's snakes can be present on any trail but are less likely to be encountered on well-traveled paths. This is one reason why State Park officials advise riders to remain on marked trails. Wear sturdy shoes, preferably boots, while riding and remain in calling range of trailmates. And take heart: Experts say humans have a greater chance of being struck by lightning than of being bitten by a snake. Besides snakes, the trail rider may occasionally meet up with an alligator. In fact, wildlife officials estimate there is probably at least one 'gator in any given Florida waterway. If trails traverse water, always keep an eye out for the alligator's protruding snout and eyes that it keeps above water while the rest of it is submerged. Presence of alligators also may be detected by their voices. Males (bull) 'gators have a heart-stopping bellowing roar. Females also

Never leave a horse inside a trailer for longer than a few minutes. Florida's sun can turn a horse trailer into an oven.

---

bellow, but with less volume and ferocity. From spring to early fall, the female alligator lays its eggs in nests of plant and grass debris. These nests are four to seven feet in diameter and should never be disturbed. Protective mother alligators can be disturbingly aggressive. Never feed alligators, even as a brave joke. Once the feast has ended, the persistant 'gator may grow peevish. Alligators can move outrageously fast. Like any lizard, the alligator raises itself up on legs that seem to sprout amazing length and strength. Alligators have been rumored to outrun a Quarter Horse in a quarter-mile, and have been seen scrambling handily over barbed wire fences. These reptiles can grow up to 20 feet long and have jaws that can snap a horse's cannon bone. Fortunately, unless they have been previously fed by earlier trail riders, alligators are shy and retiring.

Besides the other natural, occasionally hazardous trail creatures, there are birds and beasts that make ol' Dobbin stop and think twice about trail rides. Florida's rural areas abound in skunks, 'possum, raccoons, wild turkey, deer, armadillo, even the very occasional black bear. As previously mentioned, these animals avoid well-used human trailways and many are evident only at night. Horses should be acquainted with such "horrors" as quail that suddenly flush from underbrush or the random little harmless black snakes that scurry across the path. These meetings can be set up by spending some trail hours near home base.

A problem near many of Florida's home trails, however, is traffic. As the Sunshine State's popularity grows, developers gobble up riding areas. The result is trails nowadays often are buried in the recesses of public and private acreage. And to reach those hideaways the horseback rider must travel along major thoroughfares. In any major confrontation between horse and horsepower, the horse and its rider are going to be losers. Often, such accidents are "a failure to communicate." The automobile driver assumes the horse person is safely off to the side of the road and the horse rider presumes the automobile will pass quietly along its way. Unfor-

A new horse owner's common complaint is that the horse spooks at an object it just quietly passed in the other direction. This quirk is rooted in how horses view the world. Human binocular vision creates one panoramic image; horses see a monocular world with two different views coming into each eye at the same time. Although the horse has seen an object with its right eye, reversing direction means the animal is seeing that same object with its left eye for the very first time.

---

tunately for motorists, there are horses who spook into the road; sadly for horseback riders, there are motorists who take malicious delight in trying to get a rider's attention and who gun their motors or honk their horns.

Ironically, what motorists do not realize is they can be killed if their car hits a horse. An equine's 1,000-plus poundage may slide up the hood of a car and into the windshield. The driver may be crushed under that half-ton weight. And in the case of pesky motorcylists, there would not be any contest. The horse might survive an accident with a motorcycle; the cyclist most likely would not.

Florida in October 1983 passed the Horse Safety Act. It states, "Any person operating a motor vehicle shall use reasonable care when approaching or passing a person who is riding or leading an animal upon a roadway or the shoulder thereof." That means motorists should quietly slow down around horses, giving the horse the right-of-way. Inasmuch as a horse and its rider can be the big losers in auto-equine accidents, it makes sense for a rider to use utmost caution on Florida's busy roadways.

Riders should always find a back route to trails, if possible. Always use a saddle on the roadway — it can be removed once that secluded trail is reached. Use a bridle or a hackamore — not just a halter — to best control a horse. Any horse may panic, even the trail-experienced ones. Riders should know their horses — nervous horses may be more skittery, even uncontrollable in traffic.

Most important in riding along roadways is to know the type of traffic that frequents a route. Trucks, with their airhorns and hissing breaks, are a particular hazard and can frighten even the most reliable horse. If possible, horseback riders should get permission from a landowner to pass through his or her property. And horses that are skittish are always better-behaved if with friends, so ride with a calm trail buddy.

Wise riders devise contingency plans in case of emergencies. For example, a rider should plan on diving from a horse that is headed into traffic. While that sounds hard-hearted, it is easier to replace a horse than to get a second chance at life. It's always possible, too, that the runaway horse can better fend for itself without a rider on its back.

In the event a motorists becomes a particular nuisance, the rider should note the license number of the vehicle and report that driver to law enforcement agents.

LONG-DISTANCE TRAIL RIDING

An off-shoot of trail riding are endurance competitions currently being enjoyed in Florida. Among these activities are various "ride and tie" contests in which teammates alternate running and riding horseback over prescribed distances.

Similarly, charities frequently hold long-distance rides in which participants raise funds based upon how many miles they successfully cover.

Other reasons endurance riding has become popular include the health benefits that accrue to both horse and rider from such outings. In Florida an organized Distance Circuit begins in September and ends in May, before the state's summer heat revs up. The South East Distance Riders Association (SEDRA) sanctions many such rides and provides rules and awards.

Generally, the long-distance rides are divided into two categories: competitive and endurance. Competitive rides are judged, and horses must complete a course — usually 25 to 35 miles long — within a prescribed time limit. Fitness of the horses are assessed by veterinarians before and after competitive distances are completed. Beginners are advised to participate in competitive rides before going into the more demanding endurance activities. Endurance contests usually cover 50-mile courses, but some beginner endurance competitions are held over 25-mile distances. The aim of endurance riding is to finish the distance with the horse in good condition. In fact, very strict pulse and respiration limits are required and any horse that does not regain its normal vital signs within a reasonable time, or comes up lame, is disqualified.

Expert endurance and competition riders stress that participants compete only physically fit, athletic horses. Months that are traditionally cool in northern climes can be hot and humid in Florida. Obese horses have died on Florida's distance trails. Riders, likewise, should be in good condition and should be able to judge whether they and their mounts are capable of safely participating in endurance or competitive rides.

Chapter Six

Extra Assistance

A few years ago a trainer down in Maryland was doing no good whatever with a horse. It looked like it could run some, but it wouldn't win. Then he got a letter from a patient in the state insane asylum, a fellow that used to train horses. "Quit sprinting that horse," the letter said. "He wants to go a distance." Our hero picked a distance race . . . and dumped his horse in. It came home rolling. "Isn't it wonderful?" people would say. "Poor old Ed, he's off his rocker and they shut him up. But still he knows more about your horses than you do. And he's in, and you're loose."

— Joe Palmer, *This Was Racing*

Every now and then even the most well read and studious horse person needs help. The need may be as basic as wanting the address of a horse club to one as serious as requiring medical expertise for a critically ill horse.

Help at all levels abounds in Florida. Neighbors often are horse enthusiasts or owners. Veterinary offices are everywhere. The state government, cognizant of Florida's high ranking as equestrian country, has agencies in place to help horse owners. And Florida's educational institutions, particularly the University of Florida, boast some of the nation's most knowledgeable horse researchers and facilities.

Tips and Quotes! Report a stolen Florida horse to the Florida Department of Agriculture: 1-800-342-5869.

Most often called upon are the hundreds of equine veterinarians on duty in all of Florida's 67 counties. Recommendations from other horse owners are the best key to selecting an equine medical person. Likewise, as the state of Florida licenses practicing veterinarians, so the state may take that license away. Complaints about veterinarians may be submitted to the Florida Board of Veterinary Medicine, 130 N. Monroe, Tallahassee, 32301, or by calling 1-800-342-7940. Florida law may compel a veterinarian to submit to mental or physical examination when his skill and the safety of his practice is in question because of illness, drunkenness or other causes.

Florida also is one of a handful of states that has a State Horse Specialist employed by the Florida Department Agriculture. He is Robert Werstler, whose main responsibility is provide a liaison between the state horse industry and government. Werstler's office also provides information on the state's pari-mutuel horseracing industry. Inquiries should be made to Werstler at the State Department of Agriculture, Room 425, Mayo Building, Tallahassee, 32301.

Similarly, the Florida Department of Agriculture operates a network of agricultural extension agents who can be found in each county. While agents specialize in agriculture, horticulture, home economics, soil conservation and forestry, they are helpful to the horse owner who is planning to pasture one or more horses. Agents can perform soil analysis, identify poisonous plants and advise horse owners about proper fertilizing and planting. In addition, the Institute of Food and Agricultural Sciences of the University of Florida provides animal medical information via a staff of extension veterinarians who publish information and hold seminars relating to horse management.

The University of Florida, in Gainesville, treats and researches not only small animal ailments, but those of horses. The College of

Figures from the American Veterinary Medical Association reveal U. S. horse owners spend about $72 each year to doctor their steeds. Only about 13 cents is earmarked by horseowners each year to go for equine research.

Veterinary Medicine is relatively young — its first patients were admitted in 1978. Now, however, about 2,500 horses are admitted to the facility each year and approximately one-fourth of the veterinary school's faculty is directly involved in equine health care, related research or teaching.

Severe wound cases, complicated diseases such as navicular and colic, are being successfully treated at the university facility. The success rate for colic is presently about 82 percent. And such racing maladies as broken bones and lung hemorrhage are being intensively studied. Other Florida industry problems such as mare infertility also are being addressed through treatment and research.

It is neonatal foal care, however, that has garnered the university's veterinary hospital national attention. It was established in 1981 when a critically ill foal, worth a potential half-million dollars, was brought to the facility. That foal died, unfortunately, but not in vain. For it prompted the prototype intensive care unit for prematurely-born foals. The facility now saves more than half of the foals that it receives. Special attention is focused on respiratory problems, brain hemorrhaging and infections. Much of the round-the-clock care of these critically ill infant horses is provided by more than 100 community volunteers and veterinary students.

Inasmuch as Florida's horse population continues to climb, demand for treatment at the Gainesville facility also soars. Expansion of the veterinary hospital has become necessary and is currently under way with state funds, donations from the horse industry and from private horse owners. The expanded medical center will include about 50 inpatient stalls, new isolation units, laboratories, treatment rooms, a corrective shoeing facility, added surgery suites, examination areas, new radiology equipment and neurological and cardiovascular diagnosis areas. An additional sports medicine complex will include a test race track to research the impact of track surface on racing horses.

ADDRESSES

National

(General Information)

American Horse Council
1700 K St. NW
Washington, D. C. 20006

(The AHC is the national horse industry's trade association. It provides research and statistical information on the horse industry. The Council also monitors legislative activity and provides information on pertinent general issues and tax developments. Horse clubs, trainers, breeders and individual horse owners may join the Council for a fee. Up-to-date costs and membership information may be obtained by writing the Council's Washington office.)

Abuse/Health

Humane Society of the United States
2100 L St., NW
Washington, D. C. 20037

American Humane Association
5351 So. Roslyn St.
Englewood, CO 80111

American Society for the Prevention of Cruelty to Animals
441 East 92nd St.
New York, NY 10028

American Horse Protection Association, Inc.
1902 T St., NW
Washington, D. C. 20009

Morris Animal Foundation
45 Inverness Drive East
Englewood, CO 80112

Exhibiting

American Horse Shows Association
598 Madison Avenue
New York, NY 10022

United States Equestrian Team, Inc.
292 Bridge St.
South Hamilton, MA 01982

Traveling
(Overnight Stabling Directory)

Equine Travelers of America
P.O. Box 322
Arkansas City, KS 67005

Major Breed Associations

Appaloosa Horse Club, Inc.
P.O. Box 8403
Moscow, ID 83843

Arabian Horse Society
P.O. Box 85
Lebanon, OH 45036

Arabian Horse Club Registry of America, Inc.
One Executive Park
7801 E. Belleview Ave.
Englewood, CO 80110

International Buckskin Horse Association
P.O. Box 357
St. John, IN 46373

American Morgan Horse Association, Inc.
P.O. Box 1
Westmoreland, NY 13490-9990

American Paint Horse Association
P.O. Box 13486
Fort Worth, TX 76118

Paso Fino Owners and Breeders Association, Inc.
P.O. Box 600
Bowling Green, FL 33834

Pinto Horse Association of America, Inc.
910 W. Washington St.
San Diego, CA 92103

American Association of Owners and Breeders of Peruvian Paso Horses
5099 Werner Court
Oakland, CA 94602

Peruvian Paso Horse Registry of North America
P.O. Box 816 R
Guerneville, CA 95446

American Quarter Horse Association
2736 West Tenth
Amarillo, TX 79168

American Saddlebred Horse Association, Inc.
929 So. Fourth
Louisville, KY 40203

United States Trotting Association (Standardbred)
750 Michigan Ave.
Columbus, OH 43215

Tennessee Walking Horse Breeders & Exhibitors Association
P.O. Box 286
Lewisburg, TN 37091

Jockey Club (Thoroughbred)
300 Park Ave.
New York, NY 10022

Florida
(General Information)

Affiliated Horse Clubs of Florida (AHOOF)
P.O. Box 16045 Snapper Creek Branch
Miami, FL 33116
(Provides specific information about trails and horse clubs in Florida.)

Equine Science

Institute of Food and Agricultural Sciences (IFAS) Animal Science Department
210 Animal Science Building
Gainesville, FL 32611

Miscellaneous

Florida State Appaloosa Association
P.O. Box 2663
South Miami, FL 33143

Florida Barrel Racers Association
Route 1, Box 1976
Plant City, FL 33566

Florida Cowboy Association
Route 2, Box 4010
Palatka, FL 32877

Draft Horse and Mule Association
Route 2, Box 293
Alachua, FL 32615

Florida State Mounted Drill Team Association
P.O. Box 1414
Melbourne, FL 32935

Florida Driving Society
740 Morningside Drive
Englewood, FL 33533

Florida Whips (Driving)
7450 Alafia Ridge Loop
Riverview, FL 33569

Florida State Farriers Association
P.O. Box 1000
Lake Helen, FL 32744

Stadium Jumping, Inc. (Grand Prix Jumping)
P.O. Box 305
Palmetto, FL 33561

Florida Horsemanship for Handicapped
P.O. Box 14371
Gainesville, FL 32604

Florida Morgan Horse Association
4900 NW 27 Ave.
Ocala, FL 32675

Mustang & Burro Association
9312 No. 17 St.
Tampa, FL 33612

Florida Quarter Horse Association
P.O. Box 756
Nokomis, FL 33555

Florida Thoroughbred Breeders' Association
4727 NW 80 Ave.
Ocala, FL 32675

NOTE: Some addresses are subject to change when a club or organization elects new officers. Addresses of regional or county clubs and organizations are available from AHOOF.

"Most people don't realize the importance of talking to their horses. Talking to your horse can be either a tone of calmness which is relaxing to the horse, or it can be loud and harsh, which can be irritating to him." Owen Brumbaugh, Amish horse trainer

Major Horse Publications in Florida

Appaloosa World
P.O. Box 1035
Daytona Beach, FL 32019
(monthly breed magazine)

Florida Horse
P.O. Box 2106
Ocala, FL 32678
(monthly Thoroughbred racing)

The Florida Cutter
P.O. Box 1153
Ocala, FL 32670
(monthly Western cutting horse)

Florida Horseman
P.O. Box 146
Altamonte Springs, FL 32701
(monthly general interest)

Horse Country
P.O. Box 17721
Orlando, FL 32860
(monthly general interest)

Horse and Pony
6229 Virginia Lane
Seffner, FL 33584
(semi-monthly general interest)

Recommended Reading

General Information

Jacobson, Patricia and Marcia Hayes, *A Horse Around the House,* Crown Publishers, 1978

Ensminger, M. E., *Horses and Tack,* Houghton Mifflin, 1977

Stoneridge, M. A. (Ed.), *Practical Horseman's Book of Horsekeeping,* Doubleday, 1983

Price, Steven D. (Ed.), *The Whole Horse Catalog,* Simon and Schuster, 1977

Equipment

Richardson, Julie (Ed.), *Horse Tack*, William Morrow, 1981

Baker, Jennifer, *Saddlery and Horse Equipment,* Arco, 1985

Health and Grooming

Harris, Susan, *Grooming to Win,* Charles Scribner's Sons, 1977

Rossdale, Peter and Susan M. Wreford, *The Horse's Health from A to Z,* Arco Publishing Co., 1974

Hayes, Captain Horace M., *Veterinary Notes for Horse Owners,* Arco Publishing Co., 1968

Breeding

Lose, M. Phillis, *Blessed are the Brood Mares,* Macmillan, 1978

Rossdale, Peter, *Horse Breeding,* David and Charles Publishing, 1981

Driving

Ganton, Doris, *Breaking and Training the Driving Horse,* Wilshire,

Riding — Dressage

Wilde, Louise M., *Guide to Dressage,* Breakthrough Publications, 1984

Burton, Maj. Gen. J. R., *How to Ride a Winning Dressage Test,* Houghton Mifflin, 1985

Riding · English Hunt Seat and Jumping

Morris, George, *Hunter Seat Equitation,* Doubleday, 1979

Steinkraus, William, *Riding and Jumping,* Doubleday, 1969

Licart, Commandant Jean, *Start Riding Right,* Van Nostrand, 1966

Riding · English Saddle Seat

Crabtree, Helen, *Saddle Seat Equitation,* Doubleday, rev. ed. 1982

Riding - Western

Foreman, Monte, *Monte Foreman's Horse Training Science,* University of Oklahoma Press, 1986

Ball, Charles E., *Saddle Up,* Harper & Rowe, 1973

Obstacle Course Designing

Carruthers, Pamela, *Designing Courses,* Houghton Mifflin, 1978

GLOSSARY

As with any specialized occupation or career, the horse world has its own lingo. Listed below are common terms that beginners — or even old-timers — may encounter and not understand. Mastery of the language will aid even those who are toying with the idea of launching into the wonderful world of horses and need to understand catalogs and advertisements.

AHSA — abbreviation for the American Horse Shows Association, the organization that sanctions and provides rules for the top equestrian meets.

AQHA — abbreviation for the American Quarter Horse Association, which sanctions and provides rules for Quarter Horse shows.

Action — the way a horse moves. Often refers to how much animation the horse displays, especially at the trot. Terms such as "knee action" or "hock action" suggest a horse does or does not flex those joints to a great degree. See also Way of Going.

Aged — a fully mature horse. Sources vary as to what constitutes an "aged" horse. The British consider horses who are seven or eight years old to be aged. Other authorities in this country and in other nations believe an aged horse is one whose teeth have become smoothed down, which occurs at about 12 years old.

Aids — the control or direction signals given by a rider to a horse. These generally are considered, in order of importance, the rider's legs, seat, weight, hands and voice. Artificial aids are mechanical devices that assist in controlling a horse such as tie-downs, whips and crops.

Anglo-Arab — the result of crossing a Thoroughbred horse with an Arabian horse.

Anhidrosis — an ailment that causes failure of the horse to sweat. It is also called non-sweating syndrome.

Balanced seat — the harmony between horse and rider when the rider is relaxed and squarely placed in the saddle.

Balling gun — a metal or plastic tube, usually about 17 inches long that pops medication — boluses or capsules — into a horse's mouth.

Banged tail — a horse's tail that has been cut off straight across at the bottom.

Bars (hoof) — area that forms the outside cleft of the frog.

Bars (mouth) — gum area that is free of teeth in which a bit fits.

Bay — a brown horse whose mane and tail are black. The lower legs and face may also have black coloring.

Bedding — any material such as shavings or straw that provides a soft, clean place for a horse to lie down in a stail.

Bight — portion of the reins that are left over; the ends that fall from the rider's hands.

Billet strap — leather straps of the English saddle to which the girth attaches.

Bit — a controlling device, usually made of metal, that fits onto the bridle and into the horse's mouth. Bits operate on the principle that a horse responds to the rider's wishes in order to avoid pain. Thus, different styles of bits exert pressure on different areas of a horse's mouth such as the tongue, the bars (lower jaw gums), the roof of the mouth, the lips, the chin, the nose and the poll. The most common types of bits are curbs and snaffles. See also Curb; see also Snaffle.

Blacksmith — a horseshoer. Term originated from craftsmen who hand-wrought iron tools, gates, house decorations in a hot forge. In earliest days, usually worked with a firing specialist called a fireman. See also Farrier.

Blanks — pre-formed horseshoes that are finished as needed in a forge by the shoer.

Blemish — scars or defects that may detract from a horse's appearance or resale value, but do not adversely affect the animal's performance.

Bloom — that extra sheen of good health seen on the coat of a properly cared for horse.

Bolting — may refer to a runaway horse or to a horse that gulps its food up fast.

Bone — part of the horse's skeletal system. Often used when referring to the sturdiness of an animal that has "good bone."

Boot — a variety of protective devices made of rubber or leather that protect a horse if a hind hoof or ankle hits the knee or ankle of a front leg. Some horses, because of fatigue or faulty conformation, will strike the ankles of their rear legs together

Bots — tiny oval flecks — creme-colored eggs — found frequently on Florida horses' legs, chest and neck areas. The eggs are consumed by the horse when it licks these areas and can become adult bot flies.

Break — the initial training of a young horse to render it ridable. To break to halter means the horse's knowledge extends only to wearing a halter; to break to saddle ("saddle broke") means the horse will tolerate a saddle and rider, but its education is at the basic level.

Breakover — the point when a moving horse's hoof leaves the ground during a stride.

Bridle — essentially exists to hold the bit in a horse's mouth and to give a rider means of controlling the bit with reins.

Bridle path — a shaven portion of the horse's mane behind the ears that allows good, tangle-free fitting of a bridle or halter.

Bursatti — see Summer Sore.

Bute — a colloquial abbreviation for a common anti-inflammatory medication, phenylbutazone.

Butt chain — a short length of chain, usually covered with plastic, that fastens behind a horse's rump to both sides of a horse trailer doorway.

Cannon bone — the horse's lower leg bone, above the fetlock joint and below the knee.

Camped Out — a conformation fault in which the legs of a horse extend too far outward from the body.

Canter — the English riding version of the three-beat, easy gallop. The Western equivalent is the lope, which is performed more slowly than the canter. The word canter originates from "Canterbury pace" — the gait used by pilgrims on their way to the shrine of St. Thomas à Becket at Canterbury.

Cast — to subdue, lower and pin a horse to the ground with ropes, usually for surgery. Sometimes a horse is "stall cast" — accidentally pinned to the stall floor by getting a leg, or occasionally its head, caught underneath a stall partition.

Chestnut — a reddish coat color that ranges from golden to copper to brownish "liver"-colored. Chestnut horses are sometimes referred to as "sorrel." Also refers to the scabby-looking harmless calluses that grow inside a horse's leg above the knee.

Clean — free of unsoundness and blemishes.

Cob — technically, a type of small horse. More often refers to a size of bridle or halter to fit smaller-than-average horses.

Coggins test — a laboratory test used to detect Equine Infectious Anemia (swamp fever). The test was named for Dr. Leroy Coggins of Cornell University.

Cold-backed — a horse that displays sensitivity when weight is placed upon its back. Such horses may sink down under pressure or may react more adversely by rearing, bucking or lying down.

Cold-fitting — the fitting of ready-made horseshoes without use of fire.

Colic — a generic term for equine digestive upsets. Leading killer of horses.

Collection — the balanced horse that moves with the majority of its weight carried on its hindquarters.

Colt — a young male horse usually three years or younger. Thoroughbred colts are four years old or younger.

Concho — silver decoration used on Western tack.

Conformation — the particular body assemblage or body shape of a horse.

Contracted heels — heels that are too narrow and thought to be caused by poor nutrition, badly-fitting shoes, infection or heredity.

Cooling out — following exercise, walking a horse, usually by leading it, to lower its respiration and to allow sweat to dry.

Coronary band — a highly sensitive region situated where the hoof joins the pastern from which new hoof growth originates.

Cribbing — the equine vice of latching onto wood (fencing, stall walls or doors) with the teeth, arching the neck and sucking in air. Hardly ever cured. Cribbing straps (collars that fasten around a horse's throat) may temporarily subdue the habit.

Crop — sticks — not whips — that are less than three feet long. They are used to complement the rider's leg pressure, not to punish a horse.

Cross-firing — a faulty movement that occurs when a rear hoof strikes the bottom of a forefoot. See also Forging.

Cross-ties — lengths of rope that are attached to supports such as the stall wall or grooming rack wall. The ties snap onto a horse's halter and keep it from walking off while it is being groomed.

Croup — the sloping hindquarters from the highest point behind the saddle site to the start of the tail.

Curb bit — applies pressure to many areas of a horse's mouth, including the tongue, the bars, the roof of the mouth. Extra control is achieved by a curb bit because it also pressures the horse's poll and chin.

Dam — a foal's mother.

Dapple — spotted markings, usually about two inches across, on a horse's coat. Dapple grey horses have such round, dark spots all the time; other colored horses have dapples only while their new spring coat is growing in.

Disunited — see Lead.

Dock — the solid portion of the tail that contains vertebrae. A "docked" tail is one that has been amputated to a very short length. Docking is still done to draft horses and Hackneys that are driven. In the 1800s, both in the U. S. and Great Britain, it was fashionable to have horses with very short tails. On driving horses, tails were bobbed to prevent tangling with the harness.

Dress boots — knee-high English riding boots that are smooth with no decoration or lacings.

Dressage — fine riding and training in general. Specifically refers to a style of English riding in which the horse and rider perform intricate maneuvers in a prescribed pattern upon reaching certain points (marked by letters) in a dressage arena.

Dun — a horse color of yellow or reddish tan. Duns generally have a dark stripe up their backs and may have dark coloring on their lower legs.

Easy keeper — a horse that stays well-fed on little (but quality) amounts of feed.

Encephalomyelitis — a disease that causes a horse's brain and spinal cord to become inflamed. Transmitted primarily in Florida by mosquitoes. There are three varieties: Eastern, Western and Venezuelan.

Ewe-neck — a horse's neck that curves up as though attached upside down. A conformation flaw that makes collection difficult. Also called turkey-necked or goose-necked.

Ergot — small callus-like point directly behind the fetlock joint.

Farrier — a professional horseshoer, or blacksmith, often with some formal training in diagnosing and treating foot-related ailments. Can usually perform both cold and hot shoeing techniques.

Far side — the right side of a horse. See also Near side and Off side.

Fetlock — the joint that attaches the horse's cannon bone to the pastern bone.

Field boots — English riding boots that lace up at the ankle. See also Dress boots.

Filly — a young female horse usually three years or younger. Thoroughbred fillies may be four years or younger.

Firing — use of a hot iron or needles to treat leg injuries. The procedure causes scar tissue that strengthens injured areas such as tendons.

Fittings — add-on saddle necessities such as the girth, the stirrup leathers and irons.

Flat — a riding area or a show class without jumps.

Flat saddle — an English saddle with virtually no padding in the seat or kneeroll, and a level seat.

Flea-bitten — a grey horse color characterized by tiny flecks of another color, usually reddish-brown.

Flexion — the yielding or bending of certain parts of the horse's body such as the poll (where the head joins the neck) or the jaw necessary to achieve proper collection.

Floating — the rasping or filing down of sharp points on horses' teeth.

Flying change — an immediate change of canter (or lope) lead performed by a horse without going from the canter to the trot and back to the canter on the new lead. See also Lead.

Foal — a unweaned colt or filly.

Forehand — the portions of a horse's body in front of the center of gravity located just behind a horse's withers.

Forelock — the horse's mane between its ears that lies on its forehead. Also called foretop.

Forging — a faulty movement in which the toe of a hind foot strikes the bottom of a fore foot on the same side.

Founder — the common term for laminitis. See Laminitis.

Four-in-hand — a driving hitch of four horses — one team of two horses directly in front of a second pair of horses.

Frog — the hoof's blood pump that consists of a triangular-shaped wedge of elastic horn on the sole of the foot.

Gait — natural movements of various speeds such as the walk, trot (jog), canter (lope), and gallop. Other, specialized gaits such as the rack and slow gait are performed by certain horse breeds.

Gaited — refers to those horse that have been trained to perform specialized gaits such as the rack or the slow gait.

Gee — driving term for a left turn. See also Haw.

Gelding — a castrated male horse.

Gestation — pregnancy, usually 11 months in duration.

Get the gate — to be excused from a show class by the judge.

Girth — the strap that holds an English saddle onto the horse by girdling the body just behind the forelegs. May also refer to the circumference of that area of the body. Also called a surcingle or a cinch.

Goose-rump — a conformation flaw in which the horse's croup is short, sloping and narrows at the point of the butt.

Grade horse — an unregistered horse.

Green — a horse or rider that is in basic training.

Grey — a mix of white and black hairs that produce a muted black or white color. Greys range from nearly black to almost white. Occasionally, the black and white hairs also have a mix of chestnut hairs that produce a so-called rose grey.

Grullo — mousey brownish-black hide color.

Hack — to ride a horse at large, especially over roads, fields and through woods. A hunter class mostly on the flat (although one or two very small jumps may be required). A horse that is rented to riders by the hour or day.

Hackamore — a bitless bridle that exerts its pressure on the horse's nose, chin and poll. Thin hackamore nosebands called bosals that are made from cable wire are harsh and should only be used by experienced handlers.

Hand — a measure of horse height that equals four inches.

Haw — driving term for a right term. See also Gee.

Heat — the common term used when a mare enters her breeding or estrus cycle. More delicately put as "in season."

Hindquarters — region of the horse's body that lies to the rear of the rib cage.

Hock — joint that attaches the upper portion of a horse's hind leg to the cannon bone.

Hogged mane — a clipped or shaven mane; term especially refers to Western horses. See also Roached mane.

Hot walker — a motorized device to which horses are attached and are led around in a circle. Used by trainers to cool out horses after exercise and to warm up horses prior to exercise.

Hunter — type of horse used for hunting, usually a Thoroughbred type animal.

Hyperthermia — raised temperature due to over-heating, non-sweating syndrome or over-exertion; can be fatal in Florida.

Influenza — infectious disease with symptoms of coughing, inflamed throat, loss of appetite and nasal discharge.

In foal — a pregnant horse.

In hand — leading, rather than riding, a horse.

In-line trailer — a transporting vehicle in which horses stand in line rather than beside each other.

Interfering — a faulty way of moving during which any of a horse's legs strike or rub an opposite leg.

In the ribbons — placing well enough in a show class to earn a ribbon. Most shows award the six top finishers: 1st - blue; 2nd - red; 3rd - yellow; 4th - white; 5th - pink; 6th - green. Some shows hand out eight ribbons (7th - purple; 8th - brown).

Jog — slow, usually Western style, trot.

Jumper — a horse that is trained to jump obstacles. Usually refers to show, stadium, or grand prix level horses who must clear five- and six-foot hurdles.

Keg shoe — a ready-made horseshoe.

Lameness — inability to walk without limping.

Laminitis — inflammation (heat and pain) of the hoof's foot membranes that are called laminae.

Lead — the legs on the same side of the horse's body that are leading at the canter or gallop. On a circle, the fore and hind legs on the inside of the circle usually lead so the horse is balanced. As horses become better balanced, through advanced training, they become less dependent upon their inside legs for balance and may be purposely ridden on the "wrong" lead — called the "counter canter." A horse that canters or gallops with one lead in front and the opposite lead behind is traveling in a way termed "disunited."

Lifespan — length of the equine's life varies widely depending on the use of the horse and the care it receives. Some horses live well into their 20s or 30s and occasionally into their 40s. The oldest recorded horse was "Old Billy," a British work horse who lived to be 63 years old.

Liniment — any of various rubbing agents, usually liquid, that stimulate blood circulation. Liniments are used on swollen areas of the body or applied after exercise as a "brace" to prevent swelling.

Lip twitch — a restraint device used on the horse's upper lip. It causes discomfort that distracts a horse from misbehaving while being clipped or medicated.

Longe — exercising a horse on a large (15-30 feet in diameter) circle by use of a line or rope. Such training instructs a green horse to recognize verbal commands and to perform the various gaits (walk, trot, canter) in a circle.

Lope — the Western style canter, usually slower than in English riding.

Lunge — see Longe

Manure fork — a pitchforklike device with from five to 10 tines used for cleaning stalls. Recent rakelike styles of forks may have as many as 20 tines and do a thorough job of stall cleaning.

Mare — a female horse that is four years or older. Female Thoroughbreds officially become mares at age five.

Mud tail — a tail style that wraps the loose tail hairs up around the dock. A braided mud tail in which the tail is braided before wrapping around the dock is sometimes called a "stick." Mud tails are often seen on show or race horses that must perform in mud, where a loose natural tail would become dirty and weighted down by gooey dirt.

Mustang — native wild horse stock of the Western Great Plains.

Mutton-withered — a horse having low, practically nonexistent withers.

Muzzle — the end of the horse's nose, including the nostrils, lips and chin.

Navicular disease — lameness caused by inflamed fluid cavities (bursa) in the foot that prompt painful bone and tendon adhesions.

Near side — the left side of a horse; the side by which a horse is customarily mounted. The custom is supposed to have originated with soldiers whose swords were carried on the left side of their belts thus necessitating mounting a horse on the left.

Neck rein — to direct a horse's course by applying rein pressure to the neck instead of the mouth.

Nerving — the colloquial term for a neurectomy in which a nerve, usually the digital nerve near the hoof, conducts navicular pain.

Off side — the right side of a horse.

Parasite — an animal that lives off another at the host's expense; may live inside or outside the horse.

Park horse — originally referred to horses that were used to hack on the bridle paths of parks; horses that perform in Park show classes, usually Saddlebreds or versatility breeds such as the Arabian or Morgan horses.

Pastern — the horse's leg between the fetlock and the coronary band; the sloping portion of the horse's lower leg.

Pleasure horse — any horse that is used for pure enjoyment; in show classes, those horses that perform in a pleasurely, obedient manner.

Points — collectively the horse's mane, tail, muzzle and legs.

Pony — a breed of equine that measures 14.2-hands or shorter.

Poll — the sensitive area immediately behind a horse's ears.

Posting — colloquial term used when a rider rises to the trot at every other stride.

Proud cut — an incomplete castration. Male horses may behave as stallions but cannot produce offspring.

Puffs — see Windgall.

Push-button — a thoroughly well-trained horse that responds to subtle aids.

Registered — a horse whose ancestry is recorded with an official breed association.

Roached mane — shaven mane seen most often on three-gaited Saddlebred horses.

Roaring — a breathing difficulty in which a horse's vocal nerves are paralyzed and produce a loud breathy noise.

Roman-nosed — a horse whose facial profile protrudes slightly from the muzzle up to the point between the eyes.

Schooling — teaching a horse to accomplish whatever the rider desires.

Scours — diarrhea.

Seat — the portion of the saddle in which a rider sits. The type of riding performed such as huntseat or saddleseat in which a slightly different posture is used. A general term that describes how securely and/or properly a rider sits a horse.

Shy — a frightened horse's reaction to something it sees. That reaction may be running, rearing or bucking.

Sire — a foal's father.

Slab-sided — a horse with flat, unprotruding ribs that tend to make the horse feel narrow between the rider's legs.

Sound — in good health, not lame or insane.

Snaffle bit — applies pressure to the bars, tongue and lips of a horse.

Spook — see Shy.

Stakes race — a racing term referring to the type of race in which each horse's owner has put up equal nominating, eligibility and starting fees with the top three or four finishing horseowners dividing the pot.

Stallion — an uncastrated (whole) male horse that may be used to breed mares.

Standing — refers to a stallion that is available for breeding at a particular stud facility.

Star gazer — a horse that holds it head unattractively high.

Steward — show official who ensures that classes are run properly and according to rules.

Stocking up — fluid swelling in the horse's legs that disappears when the animal is exercised. Often occurs when a horse is left to stand in its stall for an extended period of time or when a horse is trailered a long distance. May indicate leg strain.

Stride — the amount of ground a horse covers at any given gait — walk, trot or canter.

Stud — the facility in which breeding stallions live. More recent usage has expanded the meaning to refer to a specific breeding stallion.

Summer sore — a tumorous lesion caused by flies laying stomach worm larvae in an existing wound.

Sway-backed — a horse whose back dips low.

Tack — equipment used with horses, specifically the saddle and bridle.

Temperament — a horse's disposition.

Thrifty — a horse that remains healthy with minimal care.

Thrush — a hoof ailment caused by filth, readily identified by foul odor and occasional oozing and sloughing-off of the frog.

Top line — the configuration of a horse's back. Also refers to a horse's ancestry on the sire's side.

Trappy — a way of moving in which the horse takes short, choppy strides; often caused by straight, upright pasterns.

War bridle — a training and restraining device that exerts pressure on a horse's poll and/or jaw. Very punishing and should be used only by experienced handlers.

Warmed up — exercising a horse either by longeing or riding so that its muscles are stretched and relaxed for more strenuous work.

Way of Going — refers to the manner with which a horse carries itself, the amount of action it uses in its joints, whether it is stiff or relaxed, whether it travels in a crooked or straight manner. See also Action.

Weanling — a foal that is no longer nursing milk from its mother but is not yet one year old.

Weaving — a stall vice in which a horse stands in one place and rocks back and forth as though in a trance. Stems from boredom, is rarely curable, and detracts from an animal's resale value.

Windgall — swelling of the bursa fluid in the fetlock joint usually caused by strain.

Withers — top of the horse's shoulders between base of the neck and the back.

Yearling — a colt or filly between ages one and two years.

Index

Marty Marth is a third-generation Floridian and nationally known horse journalist who writes for national horse magazines, for the University of Florida College of Veterinary Medicine, and produces a weekly newspaper horse column. For seven years she was publicist for the Florida Winter Equestrian Circuit, and is the only Floridian invited into membership of the International Association of Equestrian Journalists. She writes from her Willoughby Farm near the Suwannee River.